# BEYOND THE ALGORITHM:
## Lead What Machines Can't

# BEYOND THE ALGORITHM:
# Lead What Machines Can't

*Becoming an Indispensable Leader*
*When Technology Thinks Too*

## JOSEPH TOPINKA

MILL CITY PRESS

Mill City Press, Inc.
555 Winderley Pl, Suite 225
Maitland, FL 32751
407.339.4217
www.millcitypress.net

MILLCITY
PRESS

Editing by Danielle Schweich Freudenthal
Foreword by Cy Wakeman

Paperback ISBN-13: 978-0-98938-091-1
eBook ISBN-13: 978-0-98938-092-8

# Praise for Beyond the Algorithm: Lead What Machines Can't

*"Joe doesn't just teach strategy; he teaches humanity. And in a world racing toward automation, that's never been more important."*

— Cy Wakeman,
Drama Researcher and Founder of
Reality-Based Leadership

*"The AI era demands leaders who can navigate chaos without losing their humanity. Beyond the Algorithm shows how, with frameworks for building trust, developing people, and leading with accountability when everything else is shifting. With hard-earned wisdom and practical clarity, Joe Topinka maps the journey from traditional IT delivery to strategic value creation, showing CIOs how to orchestrate teams while keeping them grounded and resilient. His Rules of Engagement for reducing workplace drama and tools for business-led innovation offer field-tested guidance for leaders caught between legacy expectations and an AI future. A must-read ready to turn data into value."*

— Martha Heller, CEO,
Heller Search Associates

*"Beyond the Algorithm is a timely reminder that technology alone does not guarantee success. Joe makes it clear that real leadership in the AI era is built on trust, accountability, and human connection. At Netskope, we believe security should empower people and drive business outcomes, not just protect systems. This book is a powerful companion to that vision, putting the human side of technology front and center."*

— Mike Anderson,
Chief Digital and Information
Officer, Netskope

*"Joe delivers what every technology leader needs: a grounded, human roadmap for leading through AI and exponential change. Practical, clear, and filled with tools that matter, this book helps leaders thrive when the future feels uncertain."*

— Anne Hungate,
Founder and CEO,
Quantum Bloom

# Dedication

*To Danielle, my partner in every sense.*
*This book is for you.*

# Credits

*Written by*
Joseph R. Topinka

*Edited by*
Danielle Schweich Freudenthal

*Foreword by*
Cy Wakeman

# Table of Contents

# Foreword

**I'VE LONG BELIEVED** that the most extraordinary leaders aren't the ones with all the answers. They're the ones brave enough to look inward first, to own their impact, and to keep growing; no matter how many years they've led. That's why this book is so special.

Joe has been a living example of what it means to lead with heart and accountability. Over the years, I've watched him pour his wisdom, his experience, and, yes, his vulnerability into helping others become better versions of themselves. This book captures that beautifully.

It also does something I deeply admire: it takes ideas I've spent my career championing personal accountability, reality-based leadership, letting go of drama, and makes them real and practical for today's world. Joe doesn't just echo these concepts; he brings them to life in ways that leaders can actually use. That's a gift to all of us.

In these pages, you'll find more than frameworks and tools. You'll find gentle nudges and bold challenges; an invitation to step up, be real, and lead in ways that machines never could. Joe doesn't just teach strategy; he teaches humanity. And in a world racing toward automation, that's never been more important.

My hope is that as you turn these pages, you feel as supported and inspired as I have been by Joe. Let it be a companion on your leadership journey, a reminder that your greatest edge isn't your technical know-how; it's your courage to lead yourself first. Here's to building teams, cultures, and futures that no algorithm can replicate. Enjoy the journey.

**– Cy Wakeman,**
**Drama Researcher**

# About the Author

**JOE TOPINKA** is an award-winning CIO, executive coach, and author with a career spanning over two decades in business technology leadership. He is the founder of CIO Mentor, where he advises and coaches IT and business leaders across industries and sectors.

His first book, *IT Business Partnerships: A Field Guide*, introduced his proven framework for strengthening the role IT plays in business. He has served as a strategic advisor to Fortune 500 firms, startups, and public agencies, including work with leading cybersecurity and AI innovators such as Netskope and Agentic Labs. He is the former Board Chair, and now member Emeritus, of the Business Relationship Management (BRM) Institute.

Joe's work focuses on leadership, intentional relationships, demand shaping, team health, and building a modern tech advisory model. With *Beyond the Algorithm: Lead What Machines Can't*, he delivers a playbook grounded in lived experience, real client outcomes, and a deep belief in the power of accountable, business-minded technology leadership.

# Acknowledgments

**THIS BOOK, AND** so much of the work behind it, simply would not exist without the incredible people who have supported, shaped, and inspired me along the way.

To my wife, my truth, my partner, and the smartest person I know: your insight, encouragement, and unwavering belief in me have made this book stronger and more meaningful than it ever could have been on my own. Thank you for your love, your patience, and for always keeping me grounded in what matters most.

To Cy, whose ideas, questions, and fearless approach to leadership have stretched my thinking in all the best ways: I am grateful for your generosity in sharing your wisdom and for the example you set every day.

To the many wonderful friends, colleagues, mentors, and leaders who have shared their knowledge, experiences, and friendship over the years: each of you has left an imprint on how I see the world and what I've tried to capture in these pages.

To the clients and teams I've had the privilege of coaching and advising: thank you for trusting me to walk alongside you. Your openness and commitment to growth have kept me learning and inspired.

And finally, to you, the reader: thank you for giving your time and attention to these ideas. I hope they serve you well and help you lead with even more intention, heart, and courage. Because in the end, it's the love, trust, and care we show one another that truly sets us apart and that is the heart of leadership, far beyond anything a machine will ever touch.

# Introduction

I'VE HAD THE privilege of living through nearly every major technology shift over the last several decades. From mainframes to mobile, on-premise systems to the cloud, from the first data warehouses to today's AI explosion, I've watched our tools get faster, smarter, and more powerful by the year.

But through all of it, one thing has never changed: people make all the difference.

Throughout my career, the people around me – colleagues, mentors, and even those I was fortunate to lead – have profoundly influenced and shaped my leadership style and success. They taught me that the best strategies still rely on trust. That the boldest technology bets still rely on culture. That no matter how advanced our tools become, the quality of our leadership is what truly sets great organizations apart.

That is why I wrote this book.

I want to share the tools and concepts that I've leaned on throughout my career. They are practical, often inspired by others, and refined through real-world experience. They are designed to help you build trust, create accountability, navigate tough decisions, and bring out the best in your people.

Because at the heart of it, this is the most human thing we can do.

Leadership is a craft that takes practice, perseverance, and a willingness to keep learning. It is the daily work of shaping outcomes, stewarding relationships, and building something that lasts.

In reading these pages, you will find insights, examples, and plenty of prompts to help you reflect on how you lead. My hope is that you will see yourself here and feel inspired to keep growing, to keep investing in others, and to keep building teams within which people can do the best work of their lives.

No algorithm will ever replace that. And that is exactly where your greatest impact lies.

So, let's get started. There is important work ahead.

# PART I – The Shift:

# Rethinking Technology Leadership

I'VE BEEN A working professional through all the major shifts in the business technology landscape, from the early days of mainframes and ERP to the cloud, mobile, and now the tsunami of AI that has overtaken the market. Each shift has altered how organizations view technology, IT teams, and the role of CIOs. None has been more profound than the explosion of AI. Its sheer velocity, accessibility, and business potential have redefined what it means to lead in technology.

In this part, we'll explore how technology decisions have evolved from project completion to business impact. We'll examine how IT investments are increasingly judged by their contributions to strategic priorities, customer engagement, and competitive advantage, not just timelines and budgets.

We'll look at how technical debt undermines agility and why managing it is a corporate responsibility and a business imperative. We'll discuss enterprise risk and how cybersecurity, data integrity, and operational continuity are now front-and-center concerns for executive teams.

We'll also explore the rise of digital natives and how their expectations are reshaping workplace norms, forcing traditional IT to adapt. This is the era of the new CIO, the one who must Collaborate, Integrate, and Orchestrate across the enterprise.

And finally, we'll challenge the notion of digital transformation as a tech-first initiative. The real transformation is in educating and elevating the C-suite, thus enabling business leaders to engage meaningfully in technology conversations that shape the future of the organization.

# CHAPTER ONE

## The End of Traditional IT

I'LL NEVER FORGET the first time a business leader looked me in the eye and said, *"We don't have to go through the IT department anymore. I just signed up for what we need with a credit card."* In that moment, decades of tradition crumbled. The model I had grown up in, built around requirements gathering, solution design, build and run, suddenly felt outdated.

Not long after, I worked with an enterprise where the CIO discovered more than 200 cloud applications running across the business. HR was juggling five different recruiting tools. Finance had two competing expense systems. Marketing had nearly a dozen analytics platforms. None of them had gone through IT.

At first, this looked like chaos. But then it hit me: this wasn't an exception. This was the new normal. Employees could now procure technology as easily as ordering something online. And if IT leaders kept fighting it, we would be left behind.

For decades, IT operated in a predictable structure: define requirements, design a solution, build it, and run it. Business units submitted requests. IT delivered systems. Projects were scoped, funded, and launched on a timetable that reflected organizational control, not always with urgency.

# That Model Is Collapsing

The pace of business has accelerated and so has access to technology. Today, any employee with a credit card and a browser can procure a solution. This isn't theoretical; it's happening in every company, every industry. Shadow IT is no longer an anomaly. It is the new normal. And that is forcing CIOs and their teams to rethink their role inside the enterprise.

For years, CIOs pushed back against business units implementing technology on their own. We called it Shadow IT and treated it like a threat. In some cases, it was. Systems were purchased without proper risk review. Contracts were signed without legal input. Customer data was stored without security protocols. And IT was often left cleaning up the mess.

But here's what has changed: Shadow IT is not going away. The modern workforce, especially digital natives, expects fast, intuitive solutions. Cloud platforms and AI have made software procurement easier than ever. In many cases, business units can move faster without IT. The challenge is how we, as IT leaders, engage with it productively.

## The Shift: From Gatekeeper to Guide

This shift requires a mindset change. CIOs must stop thinking of themselves as gatekeepers and start acting as collaborators, integrators, and orchestrators.

Rather than fighting Shadow IT, we must embrace the reality of decentralized technology ownership. Our role is to help business leaders become solution owners, people who not only select platforms but also understand the full scope of responsibility that comes with it.

Procurement is the easy part. The real work starts after the solution is in place. Vendor management, data privacy, cybersecurity, compliance, integration, and lifecycle support are all responsibilities business owners must be ready to handle. Many welcome guidance, and this is where IT can step in as an advisor, not a controller.

## A New Language: Advocated Systems

To help clarify responsibilities, I use the term *advocated systems* (see Figure 1.1). These are platforms that the enterprise has agreed to support, fully and formally. They are staffed, funded, secured, and managed with intention. If a business unit wants IT to take on a system, it must go through a conversation about resources, risk, and value.

IT should ask:

- What will it take to support this system?
- What other commitments will be impacted?
- What is the strategic value to the organization?

Business units, in turn, should ask themselves:

- Do we have the capacity and expertise to manage this platform?
- Are we prepared to own security, compliance, and vendor oversight?
- Is this solution aligned with broader enterprise strategy?

This shared language turns ambiguous technology conversations into structured decisions.

**Figure 1.1** Advocated Systems Definition

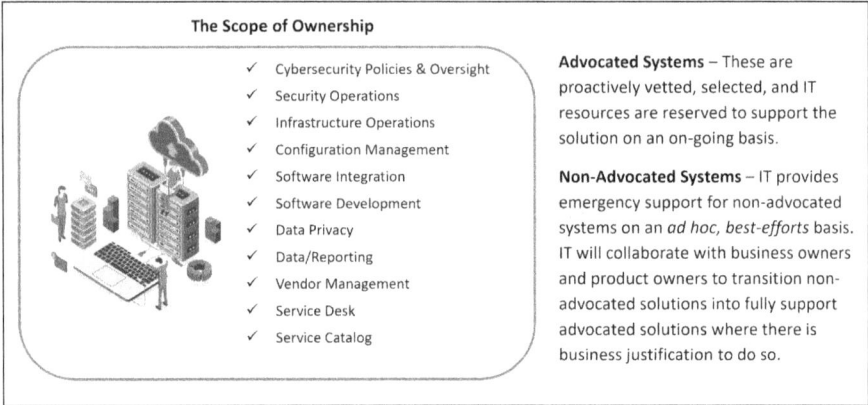

The Scope of Ownership

- ✓ Cybersecurity Policies & Oversight
- ✓ Security Operations
- ✓ Infrastructure Operations
- ✓ Configuration Management
- ✓ Software Integration
- ✓ Software Development
- ✓ Data Privacy
- ✓ Data/Reporting
- ✓ Vendor Management
- ✓ Service Desk
- ✓ Service Catalog

**Advocated Systems** – These are proactively vetted, selected, and IT resources are reserved to support the solution on an on-going basis.

**Non-Advocated Systems** – IT provides emergency support for non-advocated systems on an *ad hoc, best-efforts* basis. IT will collaborate with business owners and product owners to transition non-advocated solutions into fully support advocated solutions where there is business justification to do so.

## Three Roles: Mapping IT's Contribution

In today's landscape, IT must move beyond building and running systems. We must play multiple roles depending on the context. One useful way to visualize this is through the following three roles:

- **Advocated Systems** – IT owns and operates the platform.
- **Solution Advisor** – IT supports selection, implementation, or optimization.
- **Risk Advisor** – IT ensures that security, data privacy, compliance, and operational guardrails are in place.

By mapping each business unit or department across these roles, you can clarify where IT is engaged and where accountability sits (see Figure 1.2). This framework improves communication with business leaders and helps enterprise risk functions and boards understand who is responsible for which systems.

**Figure 1.2** A Distributed Decision Framework

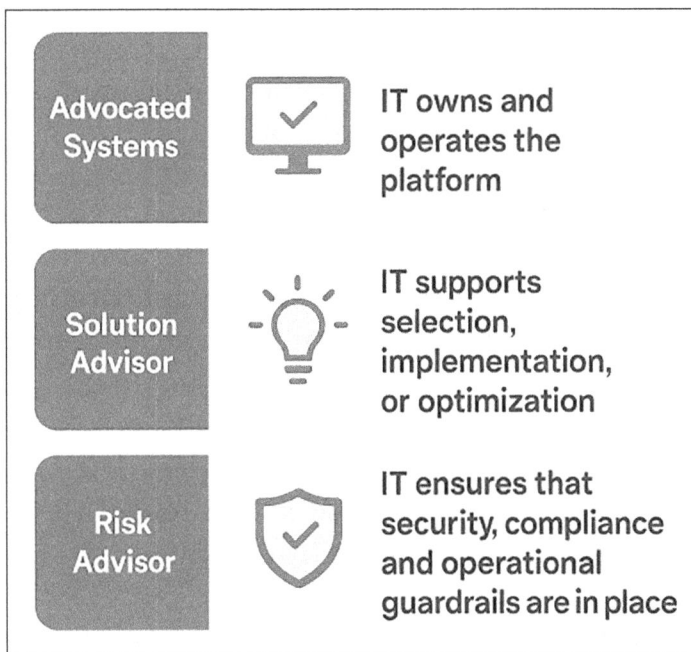

| Advocated Systems | ✓ (display icon) | IT owns and operates the platform |
| Solution Advisor | 💡 (lightbulb icon) | IT supports selection, implementation, or optimization |
| Risk Advisor | 🛡 (shield icon) | IT ensures that security, compliance and operational guardrails are in place |

As decentralized technology ownership becomes more common, organizations must create clear guardrails for how technology decisions are made and sustained. I recommend introducing a **distributed decision framework** to support this shift.

These interconnected guardrails span the **full solution lifecycle**, including:

- **Ideation** – Framing the business case and desired outcomes.

- **Vendor Selection** – Evaluating solutions with input from IT, legal, and security teams.
- **Implementation** – Ensuring data privacy, cybersecurity, and architecture alignment are built in from the start.
- **Operations** – Ongoing performance monitoring, compliance, and support.
- **Sunsetting** – Planning for secure decommissioning, data handling, and transition.

At every stage, domain experts from legal, cybersecurity, data privacy, enterprise architecture, and IT should contribute their requirements. These guardrails ensure that even as business units move independently, they are not operating recklessly. They protect data, preserve brand trust, and reinforce accountability.

Distributed decision-making combines discipline with speed by embedding expertise into a model that scales with the business (see Figure 1.3).

**Figure 1.3** Distributed Decision Guardrails

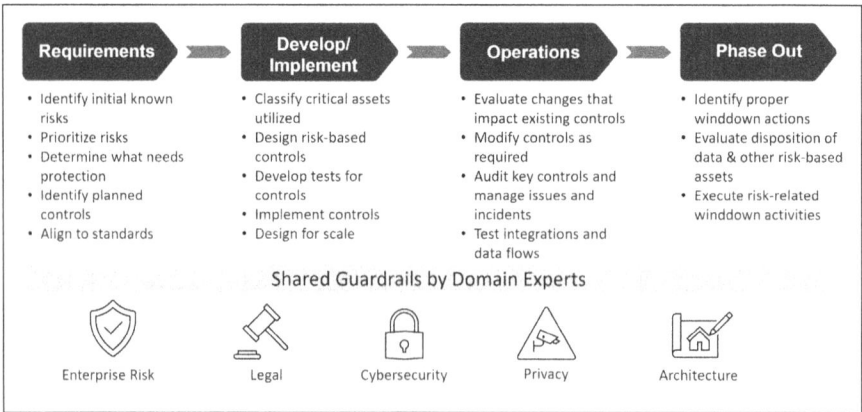

| Requirements | Develop/Implement | Operations | Phase Out |
|---|---|---|---|
| • Identify initial known risks<br>• Prioritize risks<br>• Determine what needs protection<br>• Identify planned controls<br>• Align to standards | • Classify critical assets utilized<br>• Design risk-based controls<br>• Develop tests for controls<br>• Implement controls<br>• Design for scale | • Evaluate changes that impact existing controls<br>• Modify controls as required<br>• Audit key controls and manage issues and incidents<br>• Test integrations and data flows | • Identify proper winddown actions<br>• Evaluate disposition of data & other risk-based assets<br>• Execute risk-related winddown activities |

Shared Guardrails by Domain Experts

Enterprise Risk    Legal    Cybersecurity    Privacy    Architecture

## The Real Digital Transformation

Digital transformation is often framed as a technology story. But the real transformation happens in the mindset of business leaders. True transformation comes when executives grasp how technology works and lead with the responsibility it demands, beyond simply adopting new platforms.

This mindset includes understanding technology risk, cybersecurity, integration complexity, and governance. It recognizes that owning a solution requires sustaining it through every phase of its life.

Digital transformation is the new business fluency. It demands greater accountability and a stronger partnership between business and IT.

## Embracing the New Reality

Shadow IT is a fact of modern business that invites thoughtful leadership. For CIOs, it's a powerful opportunity to build trust, shape smart technology choices, and drive the business forward. The more we educate, advise, and support our business partners, the more influence we will have. This is how IT stops being the department of "no." The best leaders learn how to say "yes" – not to everything, but to the right requests or ideas. When that "yes" is grounded in sound business judgment, it sets the tone. It clears the path for smart decisions and quietly filters out the ideas that don't hold water. Learning to say "yes" with sound judgment guides businesses toward sustainable, secure, and scalable solutions whether they're owned by IT or not.

The end of traditional IT is simply the evolution of business. It creates an opportunity to raise the bar on leadership, partnerships, and how we shape our organizations' future.

## The End of Traditional IT

The old model of IT delivery – design, build, run – is no longer compatible with the speed, access, and expectations of modern business. Shadow IT is now the default, not the exception. That shift requires IT leaders to evolve how they engage with the organization. The CIO's role is no longer to protect the gates. It is to guide, advise, and embed expertise into a distributed model of decision-making that supports speed without sacrificing discipline.

## What You Can Do Now

- **Acknowledge reality**: Accept that Shadow IT is happening and use it as a chance to engage and educate rather than resist.

- **Introduce the term *advocated systems***: Use it to clarify which platforms IT supports fully and which are owned elsewhere.

- **Map IT's contribution by role**: Create a simple matrix to define where IT is acting as owner, advisor, or risk steward.

- **Implement a distributed decision framework**: Outline the full solution lifecycle with clear guardrails for each stage, contributed by domain experts.

- **Meet business units where they are**: Don't wait for a request. Start conversations about platform ownership, accountability, and sustainability.

- **Prepare your C-suite**: Begin educating executives now on the realities of tech governance, data privacy, risk, and digital operations.

This chapter set the foundation for how IT evolves from delivery engine to strategic enabler. The rest of the book will build on these ideas where we explore the mindset and model of the modern CIO.

# CHAPTER TWO

## From Projects to Purpose

FOR DECADES, IT success was measured by a simple formula: deliver on time and on budget. This was the standard by which project managers were praised or punished. And while that discipline still matters, it no longer tells the full story.

The business landscape has changed. Customers are more demanding. Competitors are more agile. The stakes are higher. Delivering on time sets the stage, while delivering impact is the true measure of success. If projects don't improve customer journeys, create competitive advantage, or contribute to the company's strategic goals, they're just activity, not progress.

It's time to reframe how we think about success.

### Moving from Delivery to Impact

In today's environment, technology investments must be aligned with business priorities. That starts with questions such as:

- What matters most to our customers?
- Where are competitors outpacing us?
- What is the company trying to achieve?

This represents a shift in mindset that goes far beyond how we handle projects. Technology leaders must become strategy translators. Every investment, every initiative, should be traceable to an outcome the business cares about. This is the role of the **business technologist**, a leader who speaks both business and technology fluently, and connects them with purpose.

## The Problem with the Last Loudest Voice

Every company has leaders who claim to represent the voice of the customer. Some do. Others confuse their own preferences with what the market needs. These voices are often loudest in steering project decisions, and they can create noise that drowns out real priorities.

The result? IT teams are asked to chase seemingly urgent but low-impact work. Cybersecurity, privacy, architecture, and risk are often afterthoughts. Worse, the organization burns time and money on solutions that don't deliver value at the expense of other projects that could.

This is where CIOs and IT leaders must hold the line. Not necessarily to control decisions, but to inject rationality into the process. Projects must be vetted for real business impact. Ideas need to be shaped, not just accepted. Risk must be acknowledged and addressed, not ignored in the name of speed.

One way to make this easier is to build a proactive relationship with your enterprise risk team, if you have one. Framing these conversations through the lens of risk, not just IT, can make them more palatable to business leaders. If your organization doesn't have a formal enterprise risk function, you can still establish clear guardrails. Use one-on-one discussions with business leaders to introduce consistent expectations for technology decisions and emphasize the shared responsibility of protecting the organization.

A few years back, one of our regional business leaders took it upon himself to implement a customer engagement tool that he believed would revolutionize field sales. He worked directly with the vendor, paid for the pilot using discretionary funds, and rolled it out to one region without involving IT. The tool showed early promise in his small team, and word started to spread. Before long, the sales leader wanted to scale it across the organization.

That is when IT got the call. The tool had security vulnerabilities, no formal integration plan, and limited support options. More importantly, it created duplicate data flows that conflicted with our core CRM. But this executive had momentum. His pitch to senior leadership focused on anecdotal success and urgency, not long-term sustainability or risk.

We had to intervene. In hindsight, I realized I had made a mistake. I had not taken the time to build a relationship with this business unit leader. That lack of connection made the conversation about his solution harder than it needed to be. He initially saw me as someone meddling in his work rather than trying to help. Without his trust, it took longer to find alignment and move forward. That experience taught me an important lesson – strong relationships are the foundation for influence. If I hadn't done it myself, then at minimum, one of our IT business partners or Business Relationship Mangers (BRMs) should have been connected to the leader from the start.

Eventually, I sat down with him and walked through the platform's architecture, data handling, and support model. Then I showed how much effort would be needed to stabilize and scale it. The numbers were eye-opening. His enthusiasm was genuine, but his assumptions were flawed.

The key lesson? Leaders may have good intentions, but without the right questions, their ideas can become distractions. IT must lead the way in creating a structured space to evaluate value, risk, and fit – shifting the focus from the urgency of the request to whether it truly delivers strategic value.

## Technical Debt: The Hidden Cost of Misaligned Work

This is the right place to introduce **technical debt**. When organizations chase delivery at all costs, they often cut corners. Those shortcuts, temporary fixes, unsupported platforms, and overdue upgrades accumulate like interest on a loan.

Technical debt slows business, not just developers. It slows future innovation, introduces risk, and increases the cost of change.

Modern IT leaders must be proactive stewards of technical health. This includes:

- Maintaining visibility into the debt portfolio.
- Educating business partners on its impact.
- Ensuring new projects doesn't create new debt.

You don't have to eliminate all technical debt. But you must manage it intentionally. Like financial debt, some IT debt can be strategic. Too much is dangerous.

# Project Management Isn't Dead

With the rise of product thinking, agile teams, and outcome-based delivery, some have questioned whether traditional project management still has a place.

The answer: it does.

Project management remains essential for:

- Large-scale programs and enterprise-wide initiatives.
- Regulatory or risk-managed efforts.
- Budget accountability and reporting.

Project management brings structure, discipline, and clarity. Agile thinking brings speed, feedback, and iteration. These approaches are not in conflict. The best organizations blend them. They understand that structure without purpose is wasteful and that speed without guardrails is risky.

Attempting to abandon project management altogether is short-sighted. The goal is to evolve it. Equip project leaders to think in terms of outcomes; to measure value, not just milestones; and to be fluent in business conversations.

## CHAPTER SUMMARY: From Projects to Purpose

The true goal of a project is to drive the business ahead, extending past the traditional view that success was just about finishing it. IT leaders must champion this shift by:

- Aligning projects to outcomes.
- Vetting ideas with a strategic lens.
- Managing technical debt like a business risk.
- Blending agility with discipline.

Projects still matter. But what matters more is what they achieve.

## A Project Lens Versus a Purpose Lens

| Projects | Purpose-Driven |
|---|---|
| On time, on budget | Create value for customers |
| Scope and requirements focus | Outcomes and adaptability focus |
| Managed by PMO | Steered by business strategy |
| Success = delivery | Success = impact |

## What You Can Do Now

- **Start with outcomes**: Before approving or funding any initiative, ask what strategic objective it supports and how success will be measured.
- **Educate the business on technical debt**: Bring visibility to technical debt and help leaders understand its cost in terms of agility, security, and innovation.
- **Challenge the last loudest voice**: Use structured frameworks to evaluate requests, ensuring they are rooted in business value and not personal preference.
- **Blend agility with discipline**: Support agile teams, but don't discard project management. Use it where structure and coordination are essential.
- **Elevate project conversations:** Move discussions away from timelines and into impact. Track what the work is doing for the business, not just when it's done.

This chapter encouraged a shift in how we define and pursue success. Purpose, not process, is the new North Star for business technology leadership.

In the chapters ahead, we'll explore how to shape demand, prioritize investments, and lead with account-ability. But it all starts with this fundamental shift: from projects to purpose.

# CHAPTER THREE

---

# Proactive Tech Stewardship

**IT IS INCREDIBLY** tempting to ignore technical debt. If your systems seem to work and nothing has broken yet, why pour time and money into fixing what does not appear to be a problem? Many leaders fall into this trap because often we avoid what we do not fully understand. Sometimes we fall into this trap because we mistake the absence of obvious issues for proof of safety. Or we simply choose to look the other way, hoping the old saying holds true – if it's not broken, don't fix it.

Technical debt does not follow this logic. Many call it the silent killer of organizations as it quietly grows in the background, accumulating risk and complexity until the pressure becomes too much. Then the ticking bomb explodes. Headlines are filled with cautionary tales. One company lost millions almost overnight when a ransomware attack crippled its unpatched ecommerce system. That platform was responsible for 80 percent of the company's revenue. For years the system ran without incident, until suddenly it did not.

## The Hidden Burden of Technical Debt

Technical debt builds up for many reasons. Companies race to launch new products and services, layering them on top of outdated systems. Technology teams stretch limited resources, prioritizing the visible over the foundational. Older platforms continue running because replacing them seems too costly, risky, or simply not urgent.

Meanwhile, this neglected debt compounds (see Figure 3.1). Complexity increases. Costs rise. Institutional knowledge slips away as employees retire or leave. The longer technical debt goes unmanaged, the harder and more expensive it becomes to fix. And when a crisis hits – whether a cyberattack, a major system failure, or an aggressive new competitor – that hidden debt often determines just how badly the business suffers.

I experienced this firsthand when I was CIO at a large financial institution. We were nursing along an aging telephony platform that still relied on floppy disks for maintenance. Yes, floppy disks! That detail alone showed how far behind the system had fallen. I had pushed several times to replace it, but the CEO always dismissed my requests with, *"If it isn't broken, don't fix it."*

Then the inevitable happened. Just after another rejected pitch, the system crashed with no backup. My team scrambled, eventually finding a critical replacement part on eBay. We had it overnighted, installed it, and by some miracle, brought the platform back online.

The next day, the CEO asked me if I had sabotaged the system to force his hand. I hadn't, of course. But the close call became a turning point. With the CFO's support, we reframed technology upgrades as risk management, not discretionary spending. From then on, we treated technical debt as a leadership issue, one that demanded proactive stewardship rather than reactive scrambling.

Technical debt is not only a financial drain; it also erodes resilience, slows agility, and exposes serious security risks. Many technology leaders are forced to funnel precious resources into maintaining outdated systems, leaving their organizations struggling to keep up with evolving business demands.

**Figure 3.1** Compounded Neglected Debt

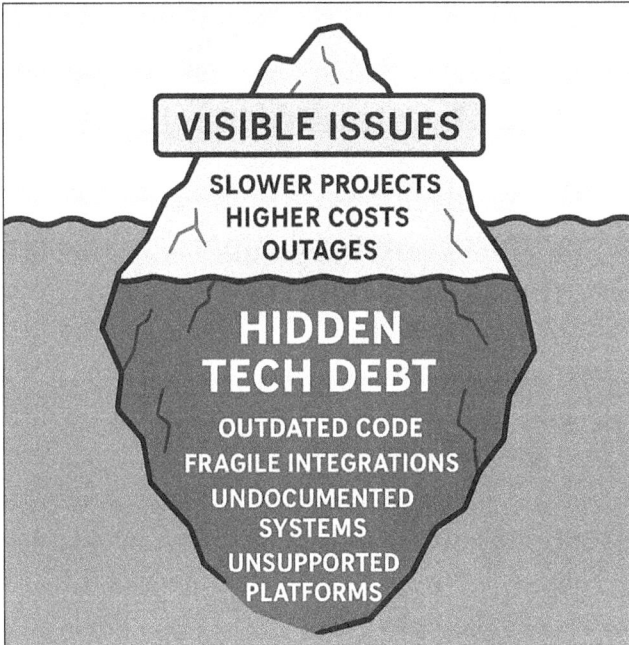

## Beyond Today's Buzz: The Evolving Tech Landscape

Technology keeps evolving. Today, the buzz is around agentic AI solutions and hyper-connected digital platforms. Tomorrow, it will be something else. The specifics change, but the pattern remains the same. Every new wave of technology promises breakthrough capabilities and introduces fresh risks.

Modern AI systems, for instance, rely on vast amounts of interconnected data, automated decisions, and learning models that can be corrupted or manipulated. If these systems sit atop a fragile foundation, the vulnerabilities multiply. Attackers look for weak points and often find them in old platforms that were never designed to handle today's speed, data volumes, or threat landscape.

This is why proactive tech stewardship is so important. Lay the groundwork now so your organization can safely and sustainably embrace whatever technology brings next.

## The Digital Native Blind Spot

Many leaders assume that as digital natives fill the workforce, technology stewardship will take care of itself. After all, these employees grew up with digital tools. They are fast adopters, unafraid of new platforms or ways of working. Being comfortable with technology, of course, is important, but being able to grasp its risks shows true technological awareness.

Many digital natives have never known a world without seamless cloud services, social platforms, and integrated third-party tools. They often take these capabilities for granted without realizing the complex security and governance structures required to keep them safe. This creates blind spots. Without deliberate education and clear policies, well-intentioned employees can expose the organization to threats they do not even see.

## A Practical Framework for Assessing Technical Debt

Recognizing technical debt is the first step. Knowing how to measure it is next. Many organizations have a vague sense they are carrying too much but struggle to quantify where the biggest risks lie.

That is why I use a structured assessment tool that examines technical debt across five core dimensions:

1. **Inventory and Infrastructure Management:** Do you maintain a comprehensive inventory of systems and understand the impact of customized legacy applications? Are you managing redundancies created by acquisitions, and ensuring products and services are actively version-controlled and not allowed to linger past practical life spans?

2. **Governance and Strategy:** Are IT purchases decentralized without adequate governance? Do you have a clear vendor consolidation strategy and standards that reduce unnecessary complexity?

3. **Product and Service Lifecycle:** Are there formal processes to evaluate and sunset outdated platforms? Do you keep products or systems alive purely because a small group still uses them, even when they are unsupported or insecure?

4. **Security and Compliance Integration:** Is cybersecurity baked into the fabric of your operations? Or is it handled as an add-on, addressed only after systems are already in place?

5. **Innovation Readiness:** Can your technology environment reliably support new initiatives without adding unacceptable risk? Or does every proposed project first have to wrestle with a legacy environment ill-suited to change?

Each area is scored based on whether it is fully managed (low risk), partially managed (moderate risk), or unmanaged (high risk). The cumulative score gives a clear picture of where the technical debt risks are most severe, guiding leaders on where to prioritize investments.

This framework does more than produce a number. It forces meaningful discussions among business and technology leaders about how their decisions today are shaping future flexibility and safety.

## Building a Culture of Stewardship

Proactive tech stewardship begins with accountability. Treat technical debt and cybersecurity as business responsibilities, not just IT problems. Stewardship requires a shift in mindset to recognize that the health of your technology foundation directly impacts your ability to grow, serve customers, and protect the brand.

It also demands collaboration. As technology decisions move into business units, the old silos among IT, cybersecurity, and operations must disappear. Everyone needs to be on the same page, making informed choices that balance speed, innovation, and long-term resilience.

Finally, stewardship calls for choosing the right partners. Few organizations can handle all this responsibility alone. Look for partners who understand your industry and regulatory landscape, have a track record of building secure and redundant platforms, and can help you modernize thoughtfully without unnecessary disruption.

## Practical Steps to Reduce Risk

Here are some straightforward actions that can help you put stewardship into practice:

- **Audit your environment:** Use the assessment framework to map out your technical debt. Identify the platforms that are unsupported, redundant, or highly customized, and document the risks they pose.

- **Prioritize cybersecurity:** Make security a board-level issue. Integrate it into every project, every vendor relationship, and every strategic conversation.

- **Educate your workforce:** Ensure that employees, especially digital natives, understand the business impact of careless technology use. This includes the risks tied to shadow IT and third-party services.

- **Build modernization into budgets:** Treat upgrades and replacement projects as strategic investments. Delaying them only makes future disruptions more likely and more expensive.

- **Foster cross-functional teams:** Break down the silos. Bring IT, business, and security leaders together to make shared decisions that consider all angles.

- **Map out a clear plan:** Create a living roadmap for modernizing systems that addresses your highest risks first. Review it regularly and adjust as threats and business needs evolve.

- **Choose partners wisely:** Work with firms that can help you modernize securely and bring specialized knowledge that complements your own team's capabilities.

# The Leadership Edge

Technical debt may be invisible on your balance sheet, but its impacts are felt everywhere. It determines how quickly you can innovate, how reliably you can serve customers, and how well you can withstand shocks. Machines can process data and algorithms can spot patterns, but only leaders can drive the decisions that keep technology healthy and secure.

By embracing proactive stewardship, you ensure your organization is not just keeping up, but also is prepared to thrive in a world where technology moves fast and risks are always evolving. This type of leadership truly goes beyond the algorithm.

## CHAPTER SUMMARY: Proactive Tech Stewardship

Technical debt is often the hidden threat that quietly undermines an organization's agility and long-term success. The debt grows quietly as companies focus on new opportunities while stacking new solutions on top of old systems. Over time, this debt weakens resilience, slows down innovation, and makes businesses more vulnerable to disruptions and cyber threats.

This chapter highlighted that while many digital natives are comfortable with technology, they must expand that comfort into a deeper understanding of its risks. It also explained that every new wave of technology, whether it is agentic AI today or the next big innovation tomorrow, depends on having a secure and healthy foundation.

A practical framework was introduced to help leaders measure technical debt across critical areas. These include understanding infrastructure inventory, improving governance and purchasing decisions, managing product and service lifecycles, integrating security, and assessing how well the environment can support future initiatives. This framework turns vague worries into clear priorities.

The core message was that machines and algorithms can handle many tasks, but only thoughtful leadership can make the decisions that keep technology healthy and secure. Proactive stewardship is essential for success in a fast-changing technology landscape where new risks are always emerging.

## What You Can Do Now

- **Map your technology landscape:** Use the assessment framework to identify which systems are outdated, unsupported, or redundant. Document where your highest risks exist.

- **Make cybersecurity part of every decision:** Treat cybersecurity as a strategic priority that is tied to the business, not just an IT issue. Include security in every discussion about new projects or partnerships.

- **Educate your teams:** Ensure that all employees, especially those who are comfortable with technology but may not understand the risks, see how their actions affect the entire organization.

- **Break down silos:** Create cross-functional teams so IT, business, and security leaders share accountability and work through decisions together.

- **Budget for modernization:** Approach upgrades and system replacements as strategic investments. Putting them off only leads to bigger problems and higher costs later.

- **Develop a living plan:** Build a roadmap for addressing technical debt and revisit it regularly to adjust for new risks and business needs.

- **Choose the right partners:** Work with firms that know your industry, understand your customers, and can guide you through modernization without unnecessary disruption.

By taking these steps now, you build a strong foundation that can handle future technology changes and the risks that come with them.

# CHAPTER FOUR

## Business-Led IT and Decentralized Innovation

TODAY, INTERNAL COMPANY organizations are no longer organized solely by departments and hierarchies. They are powered by people who span roles, solve customer problems across touchpoints, and bring technology directly into their work. Business-led IT shifts from simply buying technology to using it to spark innovation across the business. Digital natives are at the forefront, driving experiments that blend customer insight, technology, and speed. But this only works when there is a shared scaffolding, a people-powered platform that turns individual sparks into enterprise advantage. This chapter will explore how business units have become creators of technology-enabled value and why cross-functional collaboration is now the hallmark of companies that consistently meet customer needs.

### From Technology Consumers to Technology Creators

For decades, business teams relied on IT to deliver the tools they needed. They described the requirements and IT built solutions. That model was simple, but also slow and often disconnected from the real problems customers faced.

Today, that's changed. Business functions are not just asking for technology, they are creating it. Marketing teams manage sophisticated customer data platforms and automated engagement engines. Operations groups deploy IoT sensors and analytics to track assets and optimize workflows. HR leaders experiment with AI to predict turnover or personalize employee experiences.

And it goes beyond just buying software. Today, we see business units stepping into roles once reserved for IT: experimenting with technology, integrating it into daily operations, and sometimes even owning vendor relationships directly. The business units have moved from technology consumers to technology creators.

## The Digital Native Influence

Much of this momentum comes from digital natives. These employees grew up in a world of instant access, seamless interfaces, and continuous tech evolution. They are unafraid of trying new tools, running pilots, and stitching together platforms to solve problems. And unlike many IT professionals, they don't care about how technology actually works.

Their comfort with technology brings energy and fresh thinking. They look at processes and immediately imagine how apps, bots, or data flows could improve them. They spot opportunities that more seasoned leaders, weighed down by legacy constraints, might miss.

Many digital natives are comfortable with technology yet have never experienced a system crash that took days to recover from, so caution is not always second nature. They have not lived through a serious data breach or a compliance investigation. Their focus is rightly on customer and employee experience, but often without fully grasping the security, integration, or life-cycle implications.

This is where leadership comes in. The goal is not to slow them down, but to channel their drive responsibly. When supported by the right guardrails and guidance, digital natives can accelerate meaningful innovation across the enterprise.

## New Responsibilities in the Age of Business-Led IT

| New Responsibility | What It Means |
|---|---|
| Cybersecurity | Ensuring solutions meet security standards, coordinating with IT to safeguard data and systems. |
| Data Privacy | Managing how customer and business data is collected, stored, and shared, complying with privacy laws. |
| Information and Insights | Owning the quality and use of data for decision-making, reporting, and analytics. |
| Integration | Making sure new tools connect effectively with existing systems to support seamless workflows. |
| Vendor Management | Overseeing third-party providers, from selection to performance and compliance. |
| Identity and Access Management (IAM) | Controlling who has access to systems and data, and at what levels, to protect critical assets. |
| Disaster Recovery (DR) | Ensuring solutions have continuity and recovery plans in place to minimize business disruption. |
| Operational Risk | Understanding how new solutions affect business continuity, compliance, and reputational risk. |

# The People-Powered Platform

If the previous era of business focused on building technology platforms, the current era emphasizes building **people-powered platforms**: the networks of employees who work across functions, share data, and solve customer problems together.

Companies that outperform in customer experience do not organize around internal silos. They organize around journeys. They form cross-functional squads that combine marketing, sales, product, service, IT, and risk expertise. These teams look end-to-end at how customers discover, buy, use, and renew. They fix pain points collectively, not piecemeal.

This is the hallmark of decentralized innovation done well. It's not just that marketing chooses one tool, operations another, and IT yet another. It is that teams come together, pick solutions that serve the customer across touchpoints, and jointly own the outcomes. The real platform is *people*, connected by trust, common goals, and shared accountability.

## Enabling Decentralized Innovation without Chaos

All of this only works if there is a foundation that keeps speed from turning into risk. Distributed innovation needs more than encouragement; it needs clear expectations.

Cybersecurity, privacy, compliance, and architectural discipline must be embedded into every initiative, regardless of who runs it. Guardrails must exist at the ideation stage, vendor selection, implementation, and ongoing operations. They should be lightweight enough to not choke progress, but strong enough to protect the organization and its customers.

This is the new role of IT. It focuses less on building everything internally and more on enabling safe experimentation. It provides integration frameworks, shared data models, security services, and vendor oversight so business teams can innovate quickly without putting the enterprise at risk.

In this way, IT becomes an orchestrator. It pulls together domain experts in data, cyber, legal, and architecture to support business-led initiatives, making sure that as innovation scales, it does so securely and sustainably.

## Leading into the Modern CIO Era

This shift does not happen on its own. It requires intentional leadership.

Someone must keep the whole system connected, ensuring local experiments can become enterprise solutions, and protecting the brand and customers from the hidden risks of unmanaged growth. That someone is the modern CIO. No longer a gatekeeper or pure builder, the CIO is now a coach, connector, and steward of distributed innovation.

In the next chapter, we will explore what it means to lead from this new vantage point, how to influence without controlling, embed expertise without bottlenecks, and build the relationships that keep a decentralized ecosystem aligned to strategy.

# What You Can Do Now

- **Map where innovation is already decentralized:** Identify which teams or departments are selecting, building, or heavily customizing technology on their own.

- **Spot your digital native catalysts:** Engage them in conversations to learn what they are trying, where they see friction, and where they might be overlooking risks.

- **Form cross-functional squads:** Organize work around customer problems or journeys rather than departmental boundaries.

- **Reinforce guardrails:** Make sure every team knows the expectations around data privacy, cybersecurity, compliance, and architecture.

- **Celebrate shared outcomes:** Highlight wins where cross-team collaboration led to better customer or employee experiences, even if IT did not lead the project.

- **Prepare to orchestrate:** Start building the frameworks, shared services, and lightweight governance that let business-led innovations scale safely.

# CHAPTER FIVE

## Collaborate – Integrate – Orchestrate: The New CIO

IN THE PREVIOUS chapter, we explored how modern businesses are powered by people working across functions. We looked at how decentralized innovation has become essential to meeting customer needs. In business-led IT, teams do more than purchase tools swiftly. They take full ownership of solutions and outcomes. This shift has changed how companies organize, how they deliver value, and what they expect from technology.

Now we will turn to the leadership required to guide this environment. In this chapter, we'll explore how the role of the CIO is evolving into one that collaborates across business lines, integrates technology and data into a seamless fabric, and orchestrates the entire ecosystem to produce meaningful outcomes. This is the new CIO model, built on three essential responsibilities: collaborate, integrate, and orchestrate.

This model captures how work gets done in modern companies, not just in theory but in practice. Traditional hierarchies are under pressure. Customers have more influence than ever over brand perception, pricing, and how they expect to be engaged. Social media and mobile technology have shifted the power dynamic. Customer journeys do not sit inside a single department. They stretch across marketing, sales, service, delivery, operations, and support.

## The New Shape of IT: Collaborate, Integrate, Orchestrate

The old model of IT was centered on design, build, and run. For decades, that structure worked. IT gathered requirements, built systems, and kept them operating. Today, the speed of change and the demands of customers require more.

I call this new shape of IT Collaborate, Integrate, and Orchestrate. This is a straightforward way to see how the modern CIO delivers value. It is much more than just a slogan.

## Collaborate

Modern organizations solve problems through collaboration. No single team can handle customer expectations alone. Marketing understands demand and awareness. Operations knows delivery challenges. Finance sees margin pressures. IT and BRMs stand at the intersection, helping these insights come together so decisions make sense across the business.

This level of collaboration often crosses traditional department lines. Project outcome ownership is shared. It is also deeply strategic. CIOs and BRMs need to understand the landscape of cloud, data, AI, and how these tools move the business forward. They partner with business leaders to pick the right solutions and design approaches that fit strategy, not just technology trends.

At the BRM Institute, we teach how to facilitate these converged teams. In a people-powered platform, collaboration is the baseline.

# Integrate

As cloud platforms and specialized software multiply, integration has become more important than ever. Customers expect their experiences to be seamless. Employees want workflows that are connected, not fragmented. Executives need reliable data to make decisions. None of this happens on its own.

Modern IT organizations and BRMs are responsible for making sure systems, data, and processes connect properly. This includes tying internal platforms to external services, managing authentication and data flows, and maintaining clear architecture so the business can move quickly without breaking critical links.

Integration turns data into an advantage. Companies that lead in analytics design systems so insights flow to the right teams at the right time, supporting marketing, sales, product, and service in real decisions. As AI and agentic solutions become more common, the need for clean, connected data will only grow. CIOs must make sure the enterprise is prepared.

# Orchestrate

In the past, IT success meant delivering a system on time and on budget. Today, it demands more. Boards and executives expect to see the business outcomes that new technology was supposed to deliver. Did the investment reduce customer churn? Speed up product launches? Improve compliance metrics? Tracking these outcomes is now a core responsibility for IT.

Orchestration involves keeping the entire environment secure and stable. Cyber threats have become more advanced. As more customer data moves across cloud platforms and partnerships, protecting it has never been more critical. The modern CIO must coordinate delivery, ensure risk management is considered, and make security a part of every decision and every new integration.

This is a leadership function that unites technology, people, and processes to support common goals, moving well past the old model of IT operations.

## The New CIO – Collaborate-Integrate-Orchestrate

| Collaborate | Integrate | Orchestrate |
|---|---|---|
| **Cross-Functional** – Modern organizations solve problems through collaboration that crosses departments and extends to key external partners. | **Connected Systems** – As cloud platforms and software grow, integration is critical to ensure seamless customer experiences, connected workflows, and reliable executive data. | **Outcome-Focused** – Success today is measured by business outcomes, not just system delivery – reducing churn, speeding launches, and improving compliance. |
| **Strategic Alignment** – CIOs and BRMs sit at the intersection of marketing, operations, and finance, aligning decisions with insights from across the business and vendor ecosystem. | **Architected Links** – IT and BRMs ensure systems, data, and processes connect – tying internal platforms to external services, managing authentication, and designing clear architectures. | **Secure Delivery** – CIOs coordinate delivery and embed risk management, ensuring security is integral to every decision and integration. |

| Shared Ownership – At BRM Institute, collaboration is taught as foundational – sharing ownership outcomes and choosing solutions that fit strategy, not just trends. | Insight Advantage: Information Done Right – Integration turns data into advantage by making sure insights reach the right teams at the right time, supporting sales, marketing, and service – especially as AI and agentic solutions grow. | Unified Leadership – Orchestration is a leadership role that aligns technology, people, and processes to protect, enable, and achieve shared goals. |
| --- | --- | --- |

What matters now is how well your people connect these experiences. It matters whom you hire, how they work together, how quickly they can respond, and how willing they are to look beyond their own team to solve customer problems. Cross-functional squads and journey teams have replaced many old structures. This is how businesses address the complexity of modern customer expectations. This is also why roles such as BRMs have become so important. They are the connectors who help teams see the full picture, synchronize on strategy, and ensure technology decisions support outcomes that matter to customers and to the business. The people platform is the real foundation of success in our digital world.

## When This Approach Is Missing: A Real Example

Let me give you a concrete example of what happens when collaboration, integration, and orchestration are missing up front.

Years ago, I worked for an iconic footwear manufacturer and retailer. This brand had a loyal following. Customers loved their boots so much that when they started to wear out, they didn't look for replacements. They wanted the manufacturer to restore them, using the same old-world craftsmanship that made the boots famous. The company had a skilled team of cobblers ready to give these boots a new life.

The marketing team saw an opportunity to build on this loyalty. They wanted to celebrate the cobbler shop and tell stories that highlighted this unique service. They created a beautiful new website content and launched it to customers.

The campaign resonated with customers immediately. Orders for repairs and refurbishments poured in. That was the good news. The bad news was that operations was not prepared for the surge. Marketing had not coordinated closely enough with IT, operations, or customer service to make sure the shop could handle the new demand or keep customers informed about the status of their boots.

We scrambled to pull together an emergency cross-functional team. We looked at how to train more cobblers, organize the sudden flood of boots, and set up systems to give customers clear updates. Fortunately, the company had a culture that was exceptional at learning from these moments. Teams rallied quickly, protected the customer experience, and then used this event to drive a broader refresh of how customer touchpoints were managed. Customers today have more influence over brand, price, and expectations than ever before, and in this case, the company's response was celebrated on social media. The responsiveness cemented the company's reputation as a brand that truly puts customers first. It also became part of why, more than a century later, the company is are still delighting customers.

The lesson is clear. When collaboration, integration, and orchestration are not built in from the start, even the best ideas can strain operations, disappoint customers, and require costly fixes. When these elements are in place, they create resilience and deepen customer loyalty.

## Progress, Challenges, and Opportunity

Change is coming whether companies like it or not. Market forces will not slow down to protect old ways of working. Customers will continue to shape industries through their choices and demands. Digital platforms and AI will keep evolving. Some jobs will transform. Others will disappear. Resisting this reality only leads to frustration and missed opportunities.

Successful organizations are those that embrace the people-powered model. They build IT teams that thrive by collaborating across silos, integrating disconnected systems into something cohesive, and orchestrating efforts to produce real outcomes while managing risk.

The modern CIO stands at the center of this work – not as a gatekeeper, but as someone who enables others; not only as a technologist, but as a business leader who knows how to bring people, platforms, and partners together to move strategy forward.

## What You Can Do Now

- **Break down walls:** Look for opportunities where IT, marketing, operations, and finance can tackle customer problems together.

- **Invest in integration:** Avoid letting a patchwork of tools weaken customer or employee experiences. Build the systems and data backbone that supports seamless engagement.

- **Track outcomes:** Move beyond project completion metrics. Measure what your technology investments achieve for customers, revenue, and risk.

- **Elevate security:** Make cybersecurity part of every conversation. As you add more platforms, your exposure increases.

- **Support the people platform:** Develop roles such as BRMs to keep teams in sync on strategy and focused on delivering value, not just checking boxes.

- **Become the orchestrator:** Help your organization manage complexity by making sure all the moving parts, from systems to teams to external partners, work together in service of customer and business goals.

## The Leadership Edge

The old IT delivery model is fading. The CIO who thrives today is the one who collaborates across business lines, integrates systems into a unified experience, and orchestrates work so that outcomes matter and risks are managed. The focus is now on building a resilient, customer-focused, secure organization that can adapt as the world continues to change.

In the next chapters, we will explore how to deepen this leadership stance, strengthen critical relationships, and become the trusted advisor your company relies on to navigate what comes next.

| Traditional CIO | Modern CIO |
|---|---|
| Focuses on delivery: design, build, run | Focuses on value: collaborate, integrate, orchestrate |
| Measures success by projects completed on time and on budget | Measures success by business outcomes, customer impact, and managed risk |
| Operates primarily within IT boundaries | Works across the business to solve customer problems |
| Controls technology choices and implementations | Guides and enables decentralized, business-led technology decisions |
| Manages systems and uptime | Manages secure ecosystems and data-driven experiences |
| Views security as an IT responsibility | Embeds security and risk management in every business decision |
| Builds organizational silos around technology functions | Builds people platforms that connect teams and break down barriers |
| Acts as a gatekeeper | Acts as an enabler and orchestrator of strategy, technology, and talent |

# PART II – Relationships, Influence, and Accountability:

## Trust Is the Real Differentiator

*Your circumstances are not the reason you can't succeed. They are the reality in which you must succeed.*

– Cy Wakeman

IN TODAY'S TECH-DRIVEN world, the enduring value of leadership lies not in how much you know, but in how deeply you connect. This part will focus on the relational side of leadership, where trust, emotional intelligence, and personal accountability become your greatest differentiators.

You will explore how intentional relationships are built, not assumed. You will learn how to set behavioral norms that reduce drama and reinforce alignment. You will explore accountability as a driver of growth, rooted in ownership, resilience, and learning, rather than blame. And you will discover how to coach others (and yourself) using powerful tools drawn from Cy Wakeman's Reality-Based Leadership, Daniel Goleman's emotional intelligence frameworks, and stories from the field.

These next five chapters will draw from lessons learned the hard way, from emotionally expensive employees who drained energy from high-performing teams to coaching moments that rewired someone's mindset and reignited their impact. You will see how the Rules of Engagement create alignment without bureaucracy, how an emotional intelligence assessment can open up people to see their blind spots, and how self-reflection becomes a leadership superpower.

Let's dig into what it really means to lead with intent, to influence without authority, and to hold space for growth while holding people accountable to their highest potential.

# CHAPTER SIX

# Intentional Relationships

IT'S STRANGE WHEN you think about it. We spend years in school learning how to write essays, solve equations, and deliver presentations. We study business models, programming languages, and financial statements. But rarely, if ever, are we taught how to build and sustain meaningful relationships in our professional lives.

And yet, relationships are everywhere. They are at the core of every team, every customer interaction, every partnership, and every strategic decision. People are the connective tissue of business. But most companies, like most schools, treat relationship-building as something you either pick up naturally or learn the hard way. It's rarely formalized, and even less frequently coached.

I remember the moment it clicked for me. Early in my career, I had what I can only call an "aha" moment. It was one of those insights that feels both startling and obvious. I realized that the key to nearly everything I was trying to accomplish didn't lie in process or technology. It lay in people. Specifically, in the quality of my relationships with them. Influence, trust, alignment, and momentum all flow through human connection.

That realization changed the way I led, the way I partnered, and the way I measured success. It became clear that intentional relationships aren't just a nice-to-have. They are a strategic advantage.

## The Cost of a Missed Relationship

Early in my career, I was flying high. The platform I was responsible for developing and launching was gaining traction. Customers loved its flexibility and power, and I was proud of the work, the team, and the product. Most of my energy went into refining the platform, improving the architecture, and obsessing over features. What I didn't focus on though, at least not enough, was the relationship with our sales team.

Then we landed a premiere client. A big one. It should have been a moment of shared celebration. But I wasn't plugged into the sales cycle. I hadn't built meaningful relationships with the sales leaders or taken time to understand their world. So when the deal closed and the client came forward with a long list of custom requirements, I was caught off guard.

The sales team had made promises. Big ones. The client wanted new features urgently, on a timeline that, in my view, was almost impossible to deliver. I hadn't been involved in setting customer expectations. So, when it came time to deliver, I pushed back. I wanted the customer to hear the truth: where things stood and what was realistically possible.

That decision created a rift. My stance was seen as inflexible and unsupportive. The sales team felt I was jeopardizing a critical win. The tension escalated quickly, eventually reaching the highest levels of the company. In the end, I left the company feeling frustrated and unheard.

At the time, I believed I had stood for truth while the sales team had overpromised.

But something unexpected happened. I was recruited to a competitor and spent about a year away from my former situation. With that distance came clarity. I realized I had failed to do something essential. I hadn't built a trusted relationship with a key internal partner. I had been focused on the product and neglected the people. I hadn't taken the time to understand the pressures the sales team was under or how to collaborate effectively. Because I hadn't invested in the relationship, I didn't have the relationship equity to influence the outcomes when it mattered most.

Eventually, I was invited back to the company. At first, I hesitated. Could I really return after everything that had happened? I decided to come back, but with a new mindset. I made relationships a priority. I spent time listening, learning, and rebuilding trust.

Same company. Same role. Completely different outcome. This time, I wasn't just managing a platform. I was partnering with people.

## From Experience to Coaching: Helping Leaders Build Influence Through Relationships

That lesson stayed with me. It shaped how I lead and how I coach.

Today, I work with senior IT leaders who often say they want stronger relationships with the C-suite or their board, yet when I ask whether they meet one-on-one with these executives regularly, the answer is usually "no." They rely on staff meetings and governance forums. Technically, they're in the room. But relationally, they're on the outside looking in.

Once leaders reach the upper levels of their careers, they spend less time executing and more time influencing. But there's one relationship requirement that cannot be delegated: trust.

Building one-on-one relationships with executive peers and board members is essential. Yes, attending executive team and board meetings matters. But real influence grows outside those rooms in informal conversations, shared experiences, and consistent engagement. These personal connections form the foundation for collaboration and strategic alignment.

I once coached a CIO who was well-liked but struggling to move strategic initiatives forward. His interactions with the executive team were polite and professional but mostly happened in group settings. That lack of deeper connection was limiting his ability to lead.

We created a simple plan. He scheduled recurring one-on-one meetings with every member of the executive team. At first, it felt awkward. But over time, those conversations became natural and productive. He was able to float ideas informally, receive candid feedback, and align his work with what mattered most to his peers.

The results were clear. He gained support for technology investments that improved customer retention. He was no longer seen as just the IT guy. He became a trusted partner. And as he told me later, those conversations didn't just change how others saw him; they rekindled his passion for leadership.

Unfortunately, relational skills such as these are often overlooked in formal education and underdeveloped on the job. Coaches and mentors can help fill that gap. They help leaders sharpen communication, anticipate resistance, and uncover blind spots. In this case, we worked on how to frame messages, how to lead with curiosity, and how to listen without defending. We even role-played key conversations so he could walk into those meetings confident and prepared.

## Executive Relationship Inventory

Use this quick inventory to assess where you stand and where to invest next.

| Executive Stakeholder | Frequency of 1:1 Conversations | Trust Level (Low/Med/High) | Strategic Alignment | Notes/ Next Step |
|---|---|---|---|---|
| CFO | Rare | Medium | Low | Schedule lunch to better understand finance priorities. |
| CMO | Occasional | High | High | Explore ways to co-own customer analytics work. |
| CHRO | None | Unknown | Unknown | Reach out to introduce yourself and learn priorities. |
| CEO | Group only | Medium | Medium | Request 15-minute monthly check-ins or informal drop-ins. |

**Tip:** Trust grows with visibility and consistency. Start small. Ask more questions. And remember that every strong relationship is built one conversation at a time.

## The Value of Relationship Equity

Strong relationships are built over time, but they are tested during moments of tension, change, or uncertainty. That is why I coach leaders to think about **relationship equity** the same way they think about a financial account. Every interaction is either a deposit or a withdrawal. Listening without judgment is a deposit. Withholding support when your partner is under pressure is a withdrawal. Publicly celebrating someone's contribution? Deposit. Undermining them in a meeting? A hefty withdrawal.

You cannot make a meaningful withdrawal from a relationship if you have not invested first.

The most effective leaders I have worked with never wait until they need something to start building a connection. They invest early and often, through informal conversations, active listening, collaborative problem solving, and small acts of support. Over time, these investments build trust, mutual respect, and influence. That is relationship equity.

This kind of equity comes from being known, understood, and respected, rather than being liked. It gives leaders the ability to speak honestly without damaging alignment. It allows for faster collaboration when the stakes are high. And when questions arise about your team, your decisions, or even your leadership, those trusted relationships often become your first line of support. People who know you are more likely to speak up, defend your intentions, and give you the benefit of the doubt when others are taking shots or making assumptions. That is the power of relationship equity. It's what makes it possible to lead with clarity and conviction, even when others are unsure.

Looking ahead, I believe the need for this kind of equity will only grow. As AI continues to evolve, we are going to see a dramatic shift in how information is created, shared, and consumed. AI will generate videos, messages, and content that sound and look real. People will increasingly struggle to know what is authentic, what is manipulated, and what is trustworthy. In this environment, people will crave genuine connection. They will look for leaders who are grounded, consistent, and human. They will rely on trusted relationships, not algorithms, to navigate what is real and what matters.

The bottom line is this: trust and connection are not *byproducts* of leadership. They are vital *prerequisites* for it. When you invest consistently in your relationships, you build the foundation that allows you to lead through complexity, change, and opportunity.

## Pause and Reflect

Think about the key relationships in your professional life right now. Which ones are strong enough to handle a tough conversation or a disagreement? Which ones have gone quiet or feel purely transactional?

List three relationships in which you need to invest more intentionally over the next month. Then write down one simple action you can take for each: something personal, helpful, or curious that begins to build equity.

## SUMMARY: Intentional Relationships

Relationships are more than just a soft skill; they are a strategic advantage. In this chapter, we explored how failing to invest in key relationships can limit influence, create friction, and even stall your career. You saw how building trust early and often opens doors to support, alignment, and long-term collaboration. And you learned that in an increasingly complex and uncertain world, relationship equity may be the most durable asset you have as a leader.

## What You Can Do Now

- **Identify your relationship gaps:** Map your top five internal or external stakeholders and assess how strong your connection really is.

- **Make deposits before you need withdrawals:** Reach out with encouragement, support, or curiosity, not just when you need buy-in.

- **Schedule one-on-one time:** Prioritize recurring, informal conversations with peers and leaders, even if it's one short meeting per month.

- **Be consistent under pressure:** How you show up during stress or conflict tells others what kind of partner you really are.

- **Build your support network intentionally:** Remember that trusted relationships often advocate for you when you're not in the room.

# CHAPTER SEVEN

## The Accountability Framework

**PERSONAL ACCOUNTABILITY IS** the foundation of performance. It shapes how you show up every day, for your team, your peers, and your own growth. I've coached many leaders who want more influence, more opportunity, and more support. But nothing moves forward until they take real ownership of how they lead and how they respond when things do not go their way.

Cy Wakeman's definition of *accountability* has been a powerful lens for me, both personally and professionally. Blame and justifications have no role in taking accountability. Accountability centers on four simple but powerful principles: **commitment, resilience, ownership, and learning.**

These four pillars are more than leadership traits. They are daily habits. When you live them consistently, they show up in the quality of your decisions, the strength of your team, and the trust others place in you. When you ignore them, work may still look fine on the surface, but underneath, performance and morale can start to break down.

# The Four Pillars of Personal Accountability

## 1. *Commitment*

*Commitment* means being all in – no hedging, no halfway effort. When you take something on, you own it fully. It doesn't mean you have to be perfect, but you must be fully present. Commitment isn't merely saying, "I'm on board." It's proof through consistent action and accountability. It shows up in your actions, your consistency, and your willingness to follow through even when work becomes challenging. People can tell when you are truly committed. They feel it in your tone, your follow-through, and your willingness to stay engaged through uncertainty.

## 2. *Resilience*

Resilience is the ability to keep going when circumstances are difficult. For a long time, I treated the word in a fairly shallow way. I thought resilience meant pushing through or toughing it out. But over time, I learned that resilience is more strategic than that.

Planning for resilience is what good leaders do. You cannot prepare for every eventuality, but you can prepare for some. You can anticipate the friction points, think through what might go wrong, and mentally prepare for those moments. When the inevitable obstacles show up – and they almost always do – you are not surprised. You are ready. Not because you can control everything, but because you have already done some of the work in advance.

Leaders who lack resilience tend to shut down or overreact. Leaders who build resilience adjust and move forward with clarity. Conditions may change, but resilient leaders stay focused on outcomes.

## 3. *Ownership*

Ownership puts responsibility for results squarely on your shoulder – not just the effort, but the impact. This also involves staying open to feedback, even when it feels uncomfortable, or especially when it feels uncomfortable. In moments when you want to defend yourself or push back, there is often something valuable underneath. If you can listen and reflect, you have a chance to grow.

## 4. *Learning*

Learning is the fuel of long-term success. Great leaders learn from every experience, especially the hard ones. They are not just moving fast; they are improving as they go. Personal accountability involves stepping back, asking what worked and what did not, and considering how you can do better next time. Reflection is where leadership wisdom comes from.

## Personal Accountability at a Glance

| Pillar | Key Aspect | Quick Tip |
|---|---|---|
| **Commitment** | Be Fully Present | Show you're all in through actions, consistency, and staying engaged even when it's tough. |
| **Resilience** | Plan for Obstacles | Anticipate friction points and prepare mentally so you can adjust calmly when challenges come. |

| | | |
|---|---|---|
| **Ownership** | Embrace Responsibility and Impact | Stay open to feedback, especially when it's hard to hear, and focus on outcomes not excuses. |
| **Learning** | Reflect and Improve | After every effort, ask what worked, what didn't, and how you'll get better next time. |

## When To Use the Framework

I return to these four pillars often. They serve as a mirror and a reset button. I ask myself how I am showing up. I encourage the leaders I coach to do the same. One executive told me he printed the four pillars and taped them to his desk as a reminder to pause and reflect before reacting.

This framework is also the first tool I reach for when someone is stuck in blame or frustration. By shifting the focus from what happened to how they responded, we create space for growth and forward motion.

## From Drama Magnet to Steady Leader

In one organization I supported, a senior leader was responsible for a major system upgrade. It was a complex and highly visible project. He had experience, intelligence, and drive. But he also had a habit of reacting emotionally to negative feedback.

As the leader of the project management function, he heard a lot of input, some valid and helpful, but much of it noise. Instead of filtering what he heard, he took it all in and treated every comment as a potential crisis. This created tension in the team and added stress that made work harder, not better.

He did not lack ability. What he needed was a mindset shift. We worked together using the accountability framework to help him triage his day. Each week, we looked at his commitments, how he handled setbacks, what he owned, and what he had learned.

One breakthrough came when he realized how often he was reacting to uninformed opinions. He saw that he was taking the bait and letting others control his emotions. Once he saw the pattern, he began to pause. He started asking what was factual and choosing how to respond. Over time, that became more natural.

The change was visible. He became more composed. Meetings were less tense. People started seeking him out for support instead of bracing for emotional reactions. His positive influence helped the team get through a difficult migration successfully. The shift in mindset changed his leadership, and others noticed.

While he was a high performer by traditional measures, his emotional reactivity created unnecessary tension and instability. This is the kind of situation that led me to expand how I evaluate talent, not just based on **performance** and **potential**, but also on something Cy Wakeman calls **emotional expensiveness**.

*Emotional expensiveness* refers to the drag that certain individuals place on a team through drama, reactivity, or resistance to accountability. Some people look great on paper, but they consistently drain energy, time, and trust from those around them. Others may not be your top technical performers, but they elevate everyone around them. They bring calm, focus, and clarity in moments that could otherwise spiral.

# The Accountability Reset: A Weekly Habit

Once a week, take 10 to 15 minutes and ask yourself the following:

1. Where was I fully committed this week? Where was I not?
2. What setbacks did I encounter, and how did I respond?
3. What results am I proud of? What am I avoiding responsibility for?
4. What did I learn this week – about myself, my team, or my environment?

Write down your answers or talk them through with a trusted partner. Small adjustments can lead to big changes in how you show up.

## Accountability as a Mindset Shift

When I speak with groups or coach senior leaders, I often remind them that personal accountability is a choice. It is not something we inherit. It is something we learn and commit to. And when we do, it makes our roles and our impact more natural.

This mindset gives you permission to stay grounded. It helps you avoid blame and drama, and instead, respond to challenges with intention. It gives you the frame of mind to come to work ready for whatever shows up. In fact, I often think of accountability as a kind of firewall. It protects your focus and your energy from distractions, gossip, and emotional noise that can easily derail otherwise productive conversations.

This way of thinking also sets up your team for success. When you model accountability, you create a more consistent, healthy environment. It becomes easier to introduce behavioral expectations and shared norms. That is where we are headed next.

## Quick Coaching Reflection

Take five minutes and ask yourself:

- When work gets difficult, do I look for control or do I look for blame?
- Do I model the mindset I expect from others?
- Where in my day-to-day leadership could I show more commitment, resilience, ownership, or learning?

Write down one moment from the past week where you could have shown more personal accountability. What would you do differently next time?

## Looking Ahead: Rules of Engagement

Personal accountability must be the internal foundation for all of us. Teams also need structure, shared expectations, and ways to reduce friction and unnecessary drama. In the next chapter, we will explore how to create and sustain **Rules of Engagement** – the behavioral agreements that define how your team works together, especially when work becomes challenging.

## CHAPTER SUMMARY: The Accountability Framework

Accountability means not focusing on blame. It is a leadership mindset rooted in clarity, ownership, and growth. In this chapter, we explored the four pillars of personal accountability – commitment, resilience, ownership, and learning – and how they shape the way you show up as a leader.

You saw that commitment is more than just agreeing to something; it is a full-body, full-mind choice to engage with consistency. Resilience, once seen as an exercise in stoicism, is now understood as thoughtful preparation for the challenges ahead. Ownership means taking responsibility for the outcomes you produce, not just the effort you put in. And learning is the difference between repeating mistakes and evolving through experience.

Whether you are leading a team through high-pressure change or navigating daily friction points, personal accountability gives you a powerful foundation. It grounds your mindset, filters out drama, and builds trust with those around you. When practiced consistently, it sets the tone for everything that follows – including how teams align, communicate, and perform.

## What You Can Do Now

- **Use the four-pillar model as a weekly reflection tool:** Ask yourself where you are strong and where you are slipping.

- **Practice responding instead of reacting:** When tension rises, pause and ask what is factual and what your next best action is.

- **Model accountability visibly:** Let your team see how you course correct, own results, and reflect on what you are learning.

- **Use accountability as a filter:** When evaluating a situation, ask, "Am I approaching this from commitment, resilience, ownership, or learning?"

- **Recognize accountability in others:** Call it out when someone takes ownership or grows from feedback. Reinforce the behavior you want to see.

# CHAPTER EIGHT

---

# Managing Emotionally Expensive Behaviors: Keeping Accountability and Morale in Balance

## Introduction: Why This Matters

**ACCOUNTABILITY DOESN'T ONLY** mean holding people to outcomes. Authentic accountability creates an environment in which people feel safe enough to own their work, learn from mistakes, and grow. But too often, emotionally expensive behaviors undermine that kind of environment. These behaviors drain energy, stifle creativity, and quietly eat away at trust and morale.

Imagine a leader who bristles at small missteps, or a colleague who constantly second-guesses the team's abilities. Over time, these patterns create a workplace where people stop taking risks, stop speaking up, and stop investing their best ideas. Managing these moments isn't a side note to leadership. It's the very work that keeps accountability alive and well.

This chapter will lay out what these behaviors look like and provide practical language to steer tense moments toward productive outcomes. These are the human skills that keep leadership rooted in connection, where machines and metrics can't go.

# SECTION 1: Examples of Emotionally Expensive Behaviors

These behaviors commonly surface under stress or pressure. Left unchecked, they cost teams more than time or resources. They sap enthusiasm, fuel disengagement, and make people cautious and guarded.

1. ***Being overly critical and quick to criticize***
   A person who frequently points out faults without offering constructive feedback creates a tense environment. Others become hesitant to share new ideas, fearing judgment. Creativity drops, and the team spends energy managing fear instead of producing results.

2. ***Lacking trust in the team's competence***
   When someone struggles to believe the team can perform well on its own, it breeds micromanagement. This slows development, limits ownership, and signals that people aren't trusted to do their jobs.

3. ***Micromanaging***
   Constantly overseeing details robs team members of autonomy. It blocks their ability to make decisions and grow, reducing initiative and long-term capability.

4. ***Underutilizing team members' strengths***
   When individual talents are ignored or overlooked, people can't bring their full value to their work. Over time, this leads to disengagement and missed opportunities.

5. **Missing unexamined biases**
   Failing to question personal assumptions about team members' abilities often shapes unfair interactions and decisions, whether intended or not.

6. **Suppressing growth opportunities**
   When leaders fail to create chances for team members to stretch or prove themselves, it signals that growth isn't important. The team starts to feel stagnant or underappreciated.

7. **Viewing mistakes as failures**
   Seeing errors purely as failures rather than teachable moments makes people avoid taking risks. This mindset stunts development and innovation.

8. **Creating a risk-averse environment**
   If criticism or fear of mistakes is common, team members pull back. They stick with what's safe, thus limiting creative problem-solving.

9. **Failing to acknowledge effort**
   When hard work goes unrecognized, people question whether their contributions matter. Motivation and morale dip.

10. **Struggling to balance guidance and trust**
   Some leaders find it hard to give enough support without tipping into micromanagement. This imbalance either leaves teams feeling abandoned or overly controlled.

11. ***Prioritizing personal expectations over actual outcomes***
    Evaluating team performance through a narrow lens of personal preferences rather than measured improvements or agreed goals creates confusion and frustration.

All these behaviors impose a tax on team energy. They create an undercurrent of caution, lower morale, and shift focus from doing great work to managing interpersonal risk.

## SECTION 2: How to Manage These Behaviors in Real Time

Here are some practical ways to respond in the moment. Each phrase is designed to reduce defensiveness and steer the conversation toward problem-solving, without losing accountability. I've given these methods to clients when dealing with emotionally expensive people and they routinely tell me how effective they are.

### Clarifying misunderstandings

- *Phrase:* "It sounds like there might be a misunderstanding here. Can we take a moment to review the facts together?"
- *Approach:* This slows down the situation and opens space to clear up confusion without blame.

### Redirecting the conversation

- *Phrase:* "I see you're passionate about this, and it's important to get it right. How can we support the team to ensure it has what it needs?"

- *Approach:* Validates concerns, then pivots to constructive action.

### Empathizing and reframing

- *Phrase:* "I can see how this could be frustrating, especially since things aren't moving as quickly as we'd like. Let's explore what's getting in the way and how we can address it."
- *Approach:* Acknowledges feelings while guiding toward root causes and solutions.

### Offering perspective

- *Phrase:* "I understand your concerns, and it's worth considering that the team might be handling more than meets the eye. How about we check in with the team to get a clearer picture?"
- *Approach:* Invites a fuller understanding before jumping to conclusions.

### Encouraging a positive tone

- *Phrase:* "I know we all want the best outcome, and the team is working hard toward that. Let's see how we can guide people without undercutting their efforts."
- *Approach:* Reinforces the team's good intent and shifts from blame to support.

### Seeking common ground

- *Phrase:* "Let's pause to remember that we are all on the same team, aiming for the same goals. How can we work together to ensure we're aligned and moving forward effectively?"

- *Approach:* Centers shared objectives to reduce tension.

## Proposing a pause

- *Phrase:* "Maybe we should take a quick step back to reassess the situation. A fresh perspective might help us tackle this more effectively."
- *Approach:* Creates space to cool off and regroup.

## Balancing directness with diplomacy

- *Phrase:* "I appreciate your direct approach, but I'm concerned that the team might be feeling overwhelmed. How can we address this more gently?"
- *Approach:* Honors straightforwardness while nudging toward a more supportive stance.

## Addressing quick criticism

- *Phrase:* "I understand you're focused on high standards, but it feels like we're getting stuck in criticism rather than solutions. How can we shift to finding ways to support the team?"
- *Approach:* Spotlights the pattern and redirects toward problem-solving.

## The Bigger Picture

These are the human moments of leadership where accountability is either strengthened or quietly eroded. Managing emotionally expensive behaviors not only avoids conflict, it guides people into a space where growth, trust, and shared ownership can thrive. This is what machines cannot do.

## What You Can Do Now

- **Watch for stress triggers:** Watch for these behaviors in yourself. Stress and pressure can trigger them in any leader.

- **Try new language:** Practice using one or two of these phrases this week and watch how conversations shift.

- **Check in with your team:** Ask your team how supported versus micromanaged they feel. This simple question can open valuable dialogue.

- **Be intentional with words**: Recognize that small language choices have a big effect on morale, accountability, and innovation.

# CHAPTER NINE

# Rules of Engagement: Building Behavioral Norms That

## Reduce Drama and Build Trust

IN THE LAST chapter, we defined *personal accountability*, exploring how leaders can show up, take ownership, and respond to challenges with resilience and intention. This chapter will extend that thinking to the team. If accountability is how *you* show up, **Rules of Engagement** are how *we* show up together.

Every team has its own rhythm. But without clear behavioral guardrails, even talented teams struggle. Misunderstandings grow. Drama seeps in. Meetings become tense, and trust erodes. In my experience, everything hinges on how people show up for one another, which can build trust, fuel performance, and shape a healthy culture. Alignment on how people treat each other under pressure makes all the difference.

That is why I created a set of behavioral guardrails I call the **Rules of Engagement**. These are not rules for rules' sake. They are shared rules that help people better collaborate, reduce drama, and support each other consistently; especially when the stakes are high.

# Why Teams Need Behavioral Ground Rules

Early in my leadership career, I noticed that teams with lasting success had something more than technical skill. They had **unspoken trust**, and more importantly, **spoken standards** for how to treat each other. These teams did not leave behavior to chance. They defined what healthy collaboration looked like.

Most teams skip this step. They assume professional manners or expectations of mature behaviors. But when pressure hits, assumptions fall apart. People revert to personal communication habits. Tension builds. That is when real leaders introduce clarity.

The **Rules of Engagement** are how I've brought that clarity to teams. They provide a common language for how to act, react, and support one another. They help teams focus not just on what they deliver, but *how* they deliver it together.

# The 10 Rules of Engagement

These rules have been tested across teams, companies, and coaching relationships. Some were inspired by the work of Cy Wakeman. Others came from my own leadership journey, including lessons from places as unexpected as the Second City improv stage.

1. ***Embrace a Leadership Mindset.***
   Stop judging; start helping. You are not leading when you are judging. Model the mindset you want to see in others.

2. ***Get Out of Your BMW.***
   Stop **B**itching, **M**oaning, and **W**hining. Complaints without ownership drain momentum.

3. **Support It – Don't Debate It.**
Offer input during planning, but once a decision is made, don't criticize it. Rather, focus solely on how to execute. Avoid the temptation to stand around catastrophizing about why an initiative won't succeed. Undermining decisions after they are made hurts the whole team and diminishes your prowess as a leader.

4. **Identify Facts Before Stories.**
Stay grounded in what is real. Avoid dramatic assumptions or emotional narratives that create noise. Don't respond to rumors or unsubstantiated "facts," doing so flames drama and wastes energy. Encourage your team to seek the truth, ask clarifying questions, and resist the temptation to nibble on juicy drama. The story you tell yourself shapes how you show up, so make sure it is based on facts, not fiction.

5. **Be a Human API – Assume Positive Intent.**
Think of yourself as part of a human interface: an assumption of positive intent connects ideas, people, and outcomes. Just like technical APIs require clean, consistent communication protocols, so do relationships.

Assuming positive intent isn't naïve. Choose to believe that your colleagues are acting with good motives unless proven otherwise. Pause before reacting, clarify before judging, and give people a chance to explain before jumping to conclusions.

When teams make this their default posture, everything shifts. Tension fizzles out. Communication improves. Trust builds. Most conflicts don't come from malice; they come from misunderstanding. Assuming positive intent allows you to address issues constructively rather than defensively. Being a Human API involves taking responsibility for your side of the connection. Be clear, be kind, be open and expect the same in return.

6. ***Say "Yes, and...."***
This rule is inspired by the principle from improvisational theater. In improv, the phrase **"Yes, and..."** is foundational. It keeps scenes alive. It builds momentum. It invites co-creation. Saying "Yes, and..." signals that you are listening, that you accept the idea as valid, and that you are willing to build on it; even if you ultimately steer it in a different direction.

In a business context, this mindset is transformative. It replaces defensiveness with openness. It shifts conversations from "Why we can't" to "How we might." When someone presents an idea, your instinct might be to poke holes in it or explain why it won't work. That is where innovation and trust die.

Saying "Yes, and..." doesn't mean blind agreement. You do that by acknowledging the contribution and adding something that moves the idea forward. It's an inclusive way to participate in problem-solving, especially in tense or fast-paced environments.

Teams that practice this regularly become more agile, more creative, and more supportive. Over time, the culture changes. Collaboration deepens. Conversations become more constructive. People feel heard and they respond by contributing even more.

Try it in your next meeting. Notice how it changes the tone and unlocks possibility.

7. **Be Present.**
Close the laptop. Put down the phone.

*Presence* isn't defined by proximity. It's demonstrated through how fully you participate and connect. Be mentally and emotionally available to the people in front of you. In our hyper-connected world, it is tempting to multitask during meetings, scroll through messages, or check your inbox "just for a second." But the reality is that partial attention leads to partial understanding. That makes it harder to build trust or get to the root of problems.

Being present shows respect for others' time and contributions. It involves putting away devices, making eye contact, and listening without distraction. This kind of attention improves communication, sharpens focus, and sets a powerful example.

When leaders model presence, others follow. A team that values presence is more engaged, more collaborative, and more likely to surface meaningful insights that get missed in more distracted environments.

8. **Listen First.**
Many people mistake listening for waiting until it's your turn to speak. When developed, true listening becomes a skill that great leaders practice with intention. When you listen first, you create space for others to express their ideas fully. You slow down

your urge to respond, and instead focus on what is actually being said.

This kind of listening leads to better understanding, stronger decisions, and deeper relationships. It shows people that you care enough to hear their full story before jumping in with yours. Listening also helps uncover root causes, clarify expectations, and reduce unnecessary conflict.

Effective listeners ask thoughtful questions, summarize what they heard to confirm clarity, and respond with insight. Teams that build this habit develop stronger connections and more creative problem solving. Listening is one of the simplest ways to show respect – and one of the most powerful.

9. **Be Respectful.**
People think of respect as a soft skill. In reality, it is a core leadership behavior that drives performance and trust. In every interaction, you have the opportunity to either build respect or chip away at it. Respect shows up in being on time, being prepared, and honoring the contributions of your colleagues. It comes through in how you listen, avoid interrupting, stay open to ideas, and refrain from making assumptions about others' capabilities or intentions.

When teams operate with respect, psychological safety increases. People are more likely to speak up, share concerns, and get creative. They know they will be heard and treated fairly.

Respect includes holding others accountable when the standard slips. When someone talks over a colleague, disregards feedback, or breaks a commitment, respectful leaders name the behavior and reset expectations. That kind of clarity builds a healthier culture over time.

10. *Lead with Curiosity.*
Ask questions, explore context, and dig deeper before forming opinions and remember, curiosity isn't limited to people. It's critical in an era of AI and rapid technology change. Don't blindly trust data or recommendations; understand where they come from, what assumptions they carry, and how they impact your customers and teams. Curiosity builds trust, uncovers insights, and ensures decisions blend human judgment with technological power. It keeps you learning, adapting, and ahead of the curve.

## The Rules of Engagement at a Glance

| Rule | Core Idea |
|------|-----------|
| 1. **Embrace a Leadership Mindset.** | Stop judging; start helping. Model what you want to see. |
| 2. **Get Out of Your BMW.** | Stop Bitching, Moaning, and Whining. Own the solution. |
| 3. **Support It – Don't Debate It.** | Offer input early, then fully commit to the decision. |
| 4. **Identify Facts Before Stories.** | Stay grounded in reality; avoid drama and assumptions. |
| 5. **Be a Human API.** | Assume positive intent. Communicate clearly and kindly. |

| | |
|---|---|
| 6. **Say "Yes, and…."** | Grow ideas, build momentum, foster innovation. |
| 7. **Be Present.** | Close the laptop, put down the phone, give full attention. |
| 8. **Listen First.** | Understand before responding. Listening is respect. |
| 9. **Be Respectful.** | Show up prepared, honor contributions, and hold standards. |
| 10. **Lead with Curiosity.** | In the world of AI, ask questions and explore deeply. |

## How to Make the Rules Stick

When I introduced these formalized Rules of Engagement to my team, I made them visible in every all-hands meeting. I shared them on-screen and invited open conversation. At first, the silence was uncomfortable. No one wanted to go first. But eventually, one team member spoke up – and then another. Soon, people were openly reflecting on where the rules were being followed, and where we had fallen short.

What struck me was the honesty. Team members admitted where they had missed the mark and shared stories of how others helped them course correct. The conversation became routine, not awkward. Over time, the rules became part of our fabric; more than expectations, they became part of our team identity.

Other teams took notice. I was invited to share the rules with colleagues in other departments across the company. They wanted examples, not theory. They wanted to know what it looked like in practice. That was the most rewarding part: seeing teams across the business embracing the rules and making them their own.

The "Say yes, and ..." rule became a favorite. It shifted how people handled brainstorming, problem-solving, and conflict. It replaced defensiveness with forward momentum. This one idea from the improv world had a real impact on the business world.

## Practicing What You Preach

The real test came not in team meetings, but in my own career. I had joined a company as CIO, recruited by one of the founders. I moved across the country with my family to take the role. Within six months, a new CEO was brought in by the private equity firm. As is often the case, I expected changes.

The new CEO spent six months getting his bearings. Then came the restructuring. One by one, he called each executive to share his decisions. When he called me, he told me he was reducing the number of direct reports and moving my role under another leader. He admitted it might be a mistake but felt it was necessary.

In that moment, even though I felt hurt and angry, I remembered the rule: **Support it – don't debate it**. I told him I understood and that I would do everything I could to make the change a success. I didn't get into my BMW. I didn't vent. I showed up the next day and stayed committed to leading well.

The truth was that I was disappointed. Deeply. My wife got an earful at home. But I didn't let that shape my behavior at work. I focused on being the leader I wanted others to see. Six months later, the CEO called again; this time reinstating me on the executive team. He said he needed a culture carrier on the team and saw that in me.

I was honored and surprised: first, by the restructuring; then, by the reinstatement. It reminded me that people are always watching how you respond, especially during tough moments. The Rules of Engagement were not just for my team. They were for me too.

## CHAPTER SUMMARY: Rules of Engagement

In this chapter, we explored how behavioral norms shape culture and performance. The Rules of Engagement provide structure for how teams communicate, collaborate, and hold each other accountable. When the rules are visible, practiced, and reinforced, they reduce drama, build trust, and strengthen team identity.

## What You Can Do Now

- **Start the conversation:** Share the Rules of Engagement with your team. Invite discussion and ownership.

- **Focus your practice:** Choose one rule each week to focus and reflect on as a team.

- **Make it routine:** Add the rules to your team's regular meeting rhythm. Make them part of your agenda.

- **Lead by example:** Model the rules yourself; especially when you are under pressure.

# CHAPTER TEN

## Rising Above Drama: Mindset Shifts for Impactful Leadership

### The Emotional Cost of Workplace Drama

**WORKPLACE DRAMA IS** more common than most leaders would like to admit. We'd all like to think we're evolved enough to reserve drama for our personal lives and show up to work as mature professionals. The truth is that drama lurks in meetings, creeps into Slack threads, and shows up in the form of gossip, frustration, and second-guesses in the business world every day. But that drama isn't just noise; it's expensive. It consumes emotional energy, fractures trust, and derails teams. For high performers, it can be an especially sneaky trap. They believe they are doing everything correctly, yet their emotional reactions undercut their influence and effectiveness.

Over the years, I've seen this firsthand in my coaching work. Some of the most talented people I've worked with have struggled not because of technical incompetence, but because of unchecked emotional patterns. This chapter will introduce tools and frameworks that help shift your mindset, re-center on facts, and transform emotional reactivity into intentional leadership.

# Escaping the Drama Triangle (Karpman)

Even when we do our best to live by rules of engagement, we quickly slip into drama. Recognizing when we do it is half the battle. When I first learned about the Drama Triangle, a light bulb came on for me. I began to see people's behaviors more clearly, and I could identify which of the roles in the triangle they were playing, including myself when I reflected on tense moments in the office.

The Drama Triangle, developed by Dr. Stephen Karpman, outlines three types of psychological mindsets that people tend to slip into under conflict:

- **The Victim** – Feels helpless, wronged, and stuck.
- **The Rescuer** – Jumps in to help, but ends up enabling.
- **The Persecutor** – Blames, criticizes, and judges others.

These roles keep people stuck. They pull us into unproductive behavior that's driven more by emotion than reason. And that's not just metaphorical. Neuroscience tells us that when people are in drama, they're operating out of the amygdala – the emotional center of the brain. In that state, logic takes a back seat and ego runs the show.

As we like to say: your ego is not your amigo.

Recognizing the roles of the Drama Triangle is the first step to breaking free. If you're wondering whether you're stuck in one of them, consider the following:

- Do you spend a lot of time judging others?
- Do you vent more than you problem-solve?
- Do you replay stories about how you were wronged?

- Do you rescue others at the expense of your own boundaries?

Drama is natural but ineffective. But staying in drama is optional.

## The Power of the Three Questions (Cy Wakeman)

One of the most effective ways to shift out of drama and into your rational brain is to use Cy Wakeman's Three Questions. These questions are designed to move people out of their emotional response, centered in the amygdala, and into their cognitive brain, where productive problem-solving can occur.

When you feel yourself slipping into frustration, blaming, or feeling overwhelmed, ask yourself:

1. **What do I know for sure?**
   This grounds you in objective reality. It strips away assumptions, stories, and emotional overreactions.

2. **What can I do to help?**
   This puts you into action mode. It returns your focus to ownership and contribution.

3. **What would *great* look like?**
   This question lifts your mindset and sets a productive goal. It shifts your attention to creating a better outcome.

The brain cannot operate in an emotional and cognitive state at the same time. These questions act as a switch, helping you move from reaction to response, from drama to leadership.

I've used these questions with CIOs, department heads, and rising stars. They work. They are simple, memorable, and powerful tools that help people reframe what's in front of them.

## Leading with Emotional Intelligence (Daniel Goleman)

Another critical tool in the leader's toolbox is emotional intelligence. Daniel Goleman's writings on the five pillars of emotional intelligence provide a strong foundation for managing reactions, building influence, and showing up with clarity and confidence.

1. **Self-Awareness**
   Understand your emotions, strengths, and blind spots. Recognize how your mood impacts your behavior and others.

2. **Self-Regulation**
   Manage disruptive impulses. Stay calm in difficult conversations. Be adaptable.

3. **Motivation**
   Stay focused on achieving meaningful goals, not just reacting to what's in front of you. Bring energy to your work.

4. **Empathy**
   Understand others' perspectives. This is critical for trust and collaboration.

5. **Social Skill**
   Build networks, lead change, and influence without authority.

Each of these skills helps you steer clear of drama and show up as a grounded, thoughtful leader. They take practice. But when paired with the Three Questions and an awareness of the Drama Triangle, emotional intelligence becomes a lever for transformation.

## Coaching in Action – From Drama to Resilience

Whether you're a coach, leader, or teammate, one of the most powerful impacts you can have is helping others move from insight to action. After identifying a drama pattern or emotional trigger, the next step is to close the deal. Here are a few questions to use when coaching others:

- **What are you committed to?**
  Without commitment, nothing changes.

- **How will you follow through? What's your plan?**
  Make it real. Vague intentions don't create results.

- **If you face obstacles, how will you adjust?**
  This helps build resilience and keep momentum going.

In follow-up meetings, ask the following:

- **What got in the way of your commitment?**

- **What will you do differently this time?**

And when ego rears its head, ask one more question:

- **Do you want to be right, or do you want to be successful?**

That question disarms blame and redirects energy toward progress.

Here's a story that illustrates these principles in real life.

As a CIO, I had a top performer on my team with a long tenure at the company. Pat was highly regarded for her knowledge and knew our systems inside and out. Still, she was visibly frustrated. Some speculated she had once hoped to be CIO but had been passed over several times. I was just the latest in a line of leaders she had seen come and go.

We were about to embark on a major digital transformation: launching a new B2B ecommerce solution, a new ERP platform, and an updated point-of-sale system. These were long overdue changes. I had spent considerable time building trust with the executive team, understanding its priorities, and observing how customers interacted with our current systems.

From those experiences, I crafted a strategic plan that addressed critical shortcomings and opened new opportunities. After months of work, we secured funding to launch the initiative.

Pat was energized by the program and agreed to lead a significant aspect of the work. She organized her team, developed her plans, and got off to a strong start. But it didn't take long before negative stories began circulating. Some said progress was slow. Others complained that user input wasn't welcomed. The feedback was largely unsubstantiated – but it spread.

Pat reacted to everything she heard. Rather than anchoring in facts, she let the drama take root. Tension grew. She began to treat skeptics as enemies, and morale on the team suffered. I had to intervene.

Over several weeks, we worked together on her mindset. I reminded her that reacting to stories without grounding in truth is a choice. We revisited Cy Wakeman's Three Questions and practiced using them as a way to triage the day. Eventually, the situation came to a head. During one particularly honest coaching conversation, I asked her: "Why are you choosing to come to work carrying so much drama?"

It was a turning point. I invited her to let go of the emotional narrative and try a facts-before-stories mindset. I asked her to channel her frustration into resilience and action. To her credit, she took the leap. Slowly at first, but soon with growing confidence, Pat began to shift. She listened more openly, took feedback in stride, and sought moments of truth in what others shared.

The results were noticeable. Her peers took note. So did the executive team. She grew into a more effective and respected leader. Eventually, after I left the company, she stepped into the head role of IT. Her transformation was real, and it all started with a mindset shift away from drama.

## CHAPTER SUMMARY: From Drama to Leadership

Even the best leaders fall into the trap of reactivity and unproductive stories. What matters most is how quickly we recognize it, and how effectively we move ourselves and others into a more grounded, constructive mindset. Tools such as the Drama Triangle, the Three Questions, and emotional intelligence give us self-awareness and techniques to stay focused on what matters. Coaching, when used effectively, helps solidify these shifts into habits that transform leadership.

# What You Can Do Now

- **Introduce the Drama Triangle:** Use it in team development sessions. Teach your team to recognize when members are in Victim, Rescuer, or Persecutor modes.

- **Use Cy Wakeman's Three Questions:** Leverage them to coach your team (and yourself) through moments of frustration. Post them where they are visible as reminders.

- **Reflect on your triggers**: Notice when you slip into emotional reactivity and pause to reframe.

- **Coach with accountability:** Ask people what they are committed to and how they plan to make that commitment real. Follow up.

- **Reinforce emotional intelligence:** Consider it a leadership expectation. Include emotional intelligence in performance reviews, 360s, and development planning.

- **Model the mindset shift:** When you are tempted to defend, deflect, or dramatize, take a breath, ask the questions, and choose your next action from a place of clarity.

# The Takeaway

See drama for what it is, understand how it works, and choose to lead anyway.

- Workplace drama is common, but don't choose to let it define you or your culture.
- The Drama Triangle explains recurring tension patterns, recognizing the roles is the first step to change.
- Cy Wakeman's Three Questions shift you from emotion to action.
- Goleman's emotional intelligence framework deepens your ability to lead without ego.
- Coaching others to close the deal ensures that insight becomes impact.

# PART III – Architecting Value:

# Build Environments That Prioritize Impact, Innovation, and Team Health

**IN THIS PART,** we will move beyond individual leadership to the essential work of designing the structures, practices, and conditions that allow your organization to create lasting value. This is where leadership becomes architecture, where you build the systems that steer investments, shape culture, and unlock potential across the business.

We will start by exploring the CIO's pivotal role in bringing a strategic lens to every decision, moving IT from a cost center to a value architect. We will look at how to build an investment portfolio that aligns with what matters most to the business, and how to use BRMs as the connective tissue, bridging silos and acting as investment brokers who keep the organization grounded in shared priorities.

We will examine practical tools such as business-driven scoring models to ensure new initiatives get the rigorous, transparent evaluation they deserve, and project health frameworks that surface risks before they derail momentum. You'll see how a disciplined monthly investment and governance cadence keeps leaders aligned and accountable and how shining a light on staff workloads changes the conversation around what your teams can truly take on.

From there, we will explore how to structure teams and operating models that can handle everything from keeping the lights on to driving bold new bets without burning out. We will look at what it means to intentionally create space for innovation, starting with leaders getting out of their offices to see firsthand how the company actually shows up for customers and employees. And we will close by defining what great teams look like, with frameworks that foster healthy tension, clear ground rules, and a deep commitment to collective success.

This is the work of architecting value, building the scaffolding that sustains growth, drives smarter investments, and creates environments in which people and ideas can flourish.

# CHAPTER ELEVEN

## Demand Shaping with Purpose

### Elevating the CIO's Role Through Strategic Investments

**DEMAND SHAPING, EXECUTED** effectively, secures the CIO's place at the heart of business leadership. A strong demand shaping practice turns IT from a loose collection of technical projects into an intentional investment portfolio that is directly tied to the company's strategy, and ultimately, to the customers the business serves.

This is what sets apart great IT organizations. Their project portfolios are more than a list of approved initiatives. They are a reflection of how the CIO is actively driving strategic value, helping each business unit and operating function achieve its goals, all in support of delivering on the company's market promises.

### From Company Strategy to IT Execution

It starts with understanding the company's overall strategy. But that is only the beginning. CIOs and their teams also need to grasp how this strategy translates into specific priorities across every business unit and operational function. Each part of the organization has its own objectives that support the broader goals.

Even more importantly, IT must develop a strong understanding of the customers the company serves, what they value, the problems they are trying to solve, and how the company's products and services fit into their world. This outside-in perspective ensures that technology investments are focused on internal efficiencies and are aligned with creating better customer outcomes.

When the CIO operates this way, the IT portfolio becomes a deliberate expression of the company's strategic agenda. It shows how technology resources are being invested to grow revenue, improve operational efficiency, and reduce enterprise risk, all in ways that link directly back to business and customer needs.

## The Role of BRMs in Making Strategy Real

To make this happen consistently, many CIOs establish BRMs or IT Business Partner programs. These professionals are embedded within business units and operating teams. Their mandate is to deeply understand functional priorities, identify where technology can make an impact, and guide these opportunities through the investment process.

BRMs are critical connectors. They translate broad strategic themes into specific needs. They make sure IT proposals and solutions are grounded in what each part of the business is trying to achieve. Just as importantly, they bring back insights to IT leadership about emerging customer demands and competitive pressures.

Without this structure, demand shaping risks becoming an internal exercise that is disconnected from business priorities. With it, IT stops delivering mere "projects" and starts driving meaningful investments.

# Establishing IT's Role as an Investment Broker

In high-performing organizations, the CIO is the steward of enterprise investment capacity, orchestrating how limited resources are allocated to maximize strategic outcomes.

This starts with language. Talk about investments, not just projects. Maintain an investment portfolio, not an IT backlog. Frame team time as capacity allocation, not headcount.

It also requires structures that support this role. Create simple portfolio views that show where resources are currently committed. Partner with Finance so that proposals link to funding plans and benefit assumptions. Build fluency in business capabilities and refer to them explicitly when evaluating investments.

Ask these questions:

- Which business capability does this enhance or protect?
- How does this support our customer value proposition?
- Are we allocating capacity where it will make the biggest difference?

## Applying the CIO Mentor Scoring Tool

A scoring model is a critical part of demand shaping. It actively demonstrates that IT decisions are objective and grounded in strategic priorities.

| Scoring Criteria | Description (What it Measures) | Weight |
|---|---|---|
| **Strategic Fit** | How well the proposal moves in step with enterprise strategy and priorities. Does it actively advance the business at the right time and in the right way? | **30%** |
| **Operational Pain** | The extent to which the project addresses pressing business or customer problems. How much efficiency, friction removal, or process improvement will it deliver? | **15%** |
| **Cost/Benefit** | Expected financial or measurable return relative to the required investment. ROI, payback period, or other quantifiable benefits. | **20%** |
| **Risk** | How much risk does the investment mitigate or introduce? Includes tech debt and end-of-life systems, cyber/compliance exposure, vendor dependency, delivery risk, and business continuity. | **15%** |
| **Cost Certainty** | Confidence in the estimated cost, effort, and timeline. Are requirements clear, vendors known, and scope stable? | **10%** |
| **Readiness** | The organization's ability to adopt and absorb the change. Sponsorship, clarity of demand, cultural capacity, and timing. | **10%** |

Review and vet your scoring model with your executive team. This builds alignment and trust. Then use it consistently in governance sessions to keep conversations focused on shared priorities.

## Rationalizing the Current Portfolio

Before shaping new demand, assess what is already underway. Score your existing portfolio. Tag each initiative based on whether it primarily grows revenue, improves efficiency, or reduces risk.

This often reveals work that is not well connected to strategic goals or customer outcomes. Use these insights to realign, pause, or even stop initiatives that no longer fit. Your portfolio should mirror where the business is headed.

## Moving Beyond RAG: Use ProCon for Transparent Health Instead

Too many organizations still rely on the old "red-amber-green" (RAG) model to report project health. The problem is that these color codes are often based on subjective judgments, with little agreement on what triggers a red versus an amber. One leader's "amber" might be another's "green." This lack of clarity creates misunderstanding, hides risks until they become critical, and ultimately, undermines trust in how technology investments are being managed.

The following ProCon model changes all of that. It replaces guesswork with a structured approach that sets clear thresholds tied to real data on schedule, budget, scope, and team dynamics. It combines what you see, what it means, the corrective actions required, and exactly why each level is assigned. This moves your health reporting from fuzzy status updates to precise discussions about risk, tradeoffs, and which steps to take next.

# The Standard ProCon Scale for Project Health

This version works well for most technology investments, regardless of whether they are delivered through waterfall, iterative, or blended approaches. It links visible issues to defined levels of response.

| Level | What It Means | What Actions Are We Taking? | Why We Assign This ProCon? |
|---|---|---|---|
| 5 – ALL IS WELL | On track; no significant issues | – Continue execution without change.<br>– Reinforce strengths and team morale. | – Schedule on or ahead of plan.<br>– Budget at or under approved.<br>– No scope changes. |
| 4 – COUPLE OF HICCUPS, BUT MANAGEABLE | Minor impacts well within control | – Communicate in status reports.<br>– Inform Project Sponsor and PMO.<br>– Watch for trends. | – Schedule within 10% or about four weeks (large projects).<br>– Budget within 10% or $25K.<br>– Scope shifts with minor impacts. |
| 3 – HITTING SOME BUMPS | Noticeable delays or budget creep | – Notify Sponsor, CIO, PMO Director.<br>– Plan to correct in 2 weeks.<br>– Prepare change requests.<br>– Engage Finance if needed. | – Schedule 10% or more than one month behind.<br>– Budget over 10% or $25K.<br>– Scope impacts require adjustments. |
| 2 – TROUBLE AHEAD; HELP NEEDED | Major delays or cost concerns | – Hold regular meetings with PMO Director and Executive Sponsor.<br>– Keep CIO informed.<br>– Prepare formal recovery plan. | – Schedule 20% or over two months behind.<br>– Budget over 20% or $50K.<br>– Scope changes create larger risks. |

| | | | |
|---|---|---|---|
| **1 –**<br>**RED ALERT;**<br>**ACTION**<br>**REQUIRED** | Project in serious jeopardy | – Immediate review by Executive Sponsor, PMO Director, CIO.<br>– Finance revisits business case.<br>– Reassess project viability. | No credible schedule possible.<br>– Budget over 25% or $100K.<br>– Scope impacts beyond recovery. |

## The Agile-Focused ProCon Scale for Sprint-Based Delivery

For Agile teams or projects using iterative delivery, the ProCon scale emphasizes sprint goals, team dynamics, and short-cycle corrections. This keeps reviews relevant to the cadence of work.

| Level | What It Means | What Actions Are We Taking? | Why We Assign This ProCon? |
|---|---|---|---|
| **5 –**<br>**ALL IS WELL** | Sprint goals consistently met; high morale; positive feedback | – Maintain current Agile practices.<br>– Reinforce strengths. | – Sprints completed on time, 95–100% timeline adherence.<br>– 0–5% budget variance.<br>– Scope changes add clear value. |
| **4 –**<br>**COUPLE OF**<br>**HICCUPS,**<br>**BUT**<br>**MANAGEABLE** | Minor setbacks; good team collaboration | – Adjust sprint planning,<br>– Keep stakeholders informed. | – Sprints mostly on time, 85–94% adherence.<br>– 5–10% budget variance.<br>– 5–10% scope shifts with minor impacts. |

| 3 –<br>**HITTING**<br>**SOME BUMPS** | Difficulty consistently meeting sprint goals | – Conduct focused retrospectives.<br>– Reassign roles if needed.<br>– Engage stakeholders. | – Sprints facing delays, 70–84% on time.<br>– 10–15% budget variance.<br>– 10–15% scope variance. |
|---|---|---|---|
| 2 –<br>**TROUBLE**<br>**AHEAD; HELP**<br>**NEEDED** | Frequent sprint failures; low morale; concerns on viability | – Regular meetings with PMO Director, Executive Sponsor.<br>– Prepare change requests.<br>– Consider Agile coaching. | – Sprints consistently delayed, 50–69% on time.<br>– 15–25% budget overrun.<br>– 15–20% scope variance. |
| 1 –<br>**RED ALERT;**<br>**ACTION**<br>**REQUIRED** | Sprint goals rarely met; major dysfunction | – Emergency stakeholder meetings.<br>– Pause or redesign project approach; address fundamental issues. | – Less than 50% on time.<br>– Over 25% budget variance.<br>– Over 20% scope variance leading to fundamental shifts. |

## Why This Matters

A clear, documented ProCon scale gives you and your teams an objective way to describe project health. It takes emotion and politics out of the conversation. Everyone knows in advance what triggers each level, what corrective actions are required, and who needs to be involved. This transparency builds trust and keeps the focus on managing investments to deliver outcomes, not on protecting egos or hiding problems.

# Visualizing Resource Capacity

Use loading charts to show exactly where your limited capacity is being consumed. Break it out by core services, active projects, and the new demand tied to candidate projects under consideration. This approach was inspired by the CFO of a New York bank where I served as CIO. He wanted a clear view of how IT resources were being leveraged. These charts make it obvious how much capacity is truly available for new initiatives and prepare the ground for tradeoff conversations. They also demystify where IT resources are going and which services or functions are consuming them, turning vague assumptions into concrete discussions based on real constraints.

## Sample Loading Chart

| Loading Category | Month 1 | Month 2 | Month 3 |
|---|---|---|---|
| **Core Services Loading** | | | |
| Help Desk Support | 20 | 20 | 20 |
| Network and Infrastructure Ops | 15 | 15 | 15 |
| Cybersecurity Operations | 12 | 12 | 12 |
| Business Applications Support | 13 | 13 | 13 |
| **Total Core Services** | **60** | **60** | **60** |
| **Active Projects Loading** | | | |
| CRM Replacement | 8 | 6 | 4 |
| Data Warehouse Optimization | 6 | 6 | 0 |

| | | | |
|---|---|---|---|
| HR Onboarding Automation | 5 | 5 | 5 |
| Vendor Risk Tool | 7 | 7 | 7 |
| Customer Mobile App | 4 | 4 | 6 |
| **Total Active Projects** | **30** | **28** | **22** |
| **Candidate Projects Loading** | | | |
| Manufacturing IoT Pilot | 0 | 3 | 4 |
| Payroll Modernization | 2 | 2 | 3 |
| Sales Chatbot and AI | 2 | 2 | 4 |
| IT Self-Service Portal | 1 | 2 | 3 |
| ESG Reporting Automation | 2 | 3 | 3 |
| **Total Candidate Demand** | **7** | **12** | **17** |
| **Staffing Balance Summary** | | | |
| Core + Active Demand | 90 | 88 | 82 |
| Candidate Demand | 7 | 12 | 17 |
| Total Demand | 97 | 100 | 99 |
| **Shortfall or Surplus** | **-7** | **-12** | **-17** |

*Note*: This sample loading chart is for illustration only. In your actual portfolio discussions, you will also explore which skill sets are needed for each project or service area; whether taking on new initiatives requires more software engineers, help desk, or support staff downstream; and how the mix of strategic versus operational work could shift the balance of your IT organization.

# Running the Monthly Governance Rhythm

Your monthly executive portfolio meeting is where demand shaping becomes real. This is the forum where strategic priorities turn into actionable, transparent decisions on how to invest finite resources.

For many CIOs, getting the C-suite and business unit leaders into a recurring session like this feels challenging. Executives may see it as an added burden and ask:

- Why do we need to meet monthly on IT?
- Why now when we have never had to do this before?
- Isn't this just technical detail that does not require our involvement?

When I first proposed a routine portfolio investment review with the executive team, the reaction was predictable. Groans, eye rolls, and a few thinly veiled comments about "another IT meeting." The assumption was that it would be a dry, technical exercise with little value.

But once the sessions got underway, something changed. Instead of talking about servers and systems, we focused on business outcomes. We showed how investments were mapped to strategic priorities, where resources were being used, and what results those investments were delivering. The transparency mattered. Leaders could see not only the costs but also the impact their choices were having on the business.

Over time, those meetings stopped feeling like a chore. Executives began to look forward to them. They appreciated the clarity, the discipline, and the ability to shape priorities with full visibility into tradeoffs. The new process built trust because it was anchored in outcomes that mattered. The impact spread beyond IT. Other functions took notice and began adapting the approach for their own teams. What started as a dreaded IT ritual became a model for transparent, business-driven decision-making across the enterprise.

This is where demand shaping is your strongest argument. Bringing the C-suite into this process gives executives a firsthand look at how IT investments are driven by strategic alignment and business value, not technical complexity.

## Making It Tangible: What They Actually See

- The scoring model shows exactly how proposals are evaluated based on business strategy, value, and risk. It breaks down the myth that IT decisions happen in a technical black box.
- ProCon views replace fuzzy status reports with metrics tied to cost, time, and tangible outcomes.
- Resource loading charts make tradeoffs visible by showing exactly where capacity is already tied up.

These tools turn the meeting into an investment review, not a technology update.

## Showcasing Wins and Reinforcing Value

Over time, these sessions become the perfect place to showcase wins. You can highlight how a system implementation increased sales, reduced churn, or cut process times in half. This conditions the organization to look for value in every IT initiative and makes your impact clear.

Many executives who resisted these sessions in the past often become the strongest supporters. They see firsthand how this process ties together strategy, investment, and results.

## Tailoring the Investment Lens to the Business

A scoring model should always reflect what matters most to your organization. It is your documented lens for deciding why one initiative moves forward while another waits. Here are four examples that show how companies adapt their scoring to match different priorities.

## Standard Enterprise Investment Scoring

This model fits companies that balance growth, efficiency, and risk.

| Criteria | Weight | Description |
|---|---|---|
| Strategic Fit | 30% | Alignment with enterprise goals. |
| Operational Fit | 15% | Addresses known efficiency or reliability gaps. |
| End-of-Life Considerations | 15% | Replaces unsupported or aging systems. |
| Cost and Benefit | 20% | Expected ROI or business impact. |
| Project Risk | 10% | Complexity and newness. |
| Cost Certainty | 10% | Confidence in estimates. |

## AI and Digital Transformation Scoring

For companies prioritizing innovation, learning, and competitive edge through data.

| Criteria | Weight | Description |
| --- | --- | --- |
| Strategic Fit (AI Roadmap) | 25% | Drives long-term digital strategy. |
| Operational Fit | 15% | Improves workflows or automation. |
| Innovation and Learning | 15% | Builds new capabilities. |
| Cost and Benefit | 15% | Long-term value from data and automation. |
| Data Ethics and Compliance | 15% | Ensures regulatory and ethical safeguards. |
| Project Risk | 10% | AI uncertainty, data bias. |
| Technical Feasibility and Scale | 5% | Ready to implement and scale. |

## Cybersecurity-Heavy Scoring

For businesses in which compliance and threat defense are top priorities.

| Criteria | Weight | Description |
| --- | --- | --- |
| Alignment to Security Goals | 20% | Matches security and business priorities. |
| Addresses Key Security Issues | 10% | Resolves known vulnerabilities. |
| Response to Threat Landscape | 10% | Adapts to external risks. |
| Cost and Benefit | 20% | Value in losses prevented. |
| Compliance and Legal | 15% | Meets laws and standards. |

| | | |
|---|---|---|
| End-of-Life Risks | 5% | Retires risky legacy systems. |
| Project Risk | 5% | Complexity and uncertainty. |
| Cost and Time Certainty | 5% | Stability of estimates. |
| Business Continuity Impact | 5% | Operational resilience. |
| Scalability and Future Proofing | 5% | Evolves with the business. |

## Not-for-Profit and Public Sector Scoring

For organizations that prioritize mission outcomes and public value.

| Criteria | Weight | Description |
|---|---|---|
| Mission Alignment | 30% | Advances legislative or community goals. |
| Operational Impact | 15% | Improves efficiency and service. |
| Technology Modernization | 15% | Updates aging systems. |
| Public Value | 20% | Citizen impact, equity, transparency. |
| Implementation Feasibility | 10% | Community, political, or regulatory fit. |
| Financial Certainty | 10% | Confidence in costs. |
| Social Equity and Sustainability | Optional | Supports inclusivity and climate goals. |

# What You Can Do Now

Demand shaping only works if you put it into action. Here's how you can start.

- **Reframe your language and role:** Use terms such as *investment*, *portfolio*, and *capacity* in every discussion. It shapes how others see IT.

- **Map the connections:** Document how company strategy links to business unit priorities and then to IT initiatives. Show that your portfolio is delivering on the plan.

- **Strengthen BRM ties:** If you do not have formal business partners, start small by embedding IT leaders with business units to learn their priorities and shape demand.

- **Tailor your scoring tool:** Adapt your scoring model to reflect what matters most to your organization. Document it and use it consistently.

- **Visualize your capacity:** Build a simple loading chart that shows where your people are allocated. Use it to drive honest tradeoff conversations.

- **Launch your governance rhythm:** Get executive leaders together monthly to review project health, new proposals, and capacity. This becomes the heartbeat of alignment.

- **Celebrate wins:** Use these meetings to highlight how IT investments have paid off. Over time, your organization will look for value in every project.

- **Keep evolving:** As your business changes, update your scoring, your capacity models, and your BRM engagements. Demand shaping is a living discipline.

# CHAPTER TWELVE

# Structuring for Success: Teams, Talent, and Technology

**TRADITIONALLY, IT ORGANIZATIONS** structured their work into clear pathways: support and operations, business process enhancements, analytics and reporting, strategic project development, infrastructure management, and security. These streams reflected a predictable environment in which technology was largely static and under direct human control.

That landscape is evolving. As organizations adopt more data-driven automation and intelligent decision systems, each of these work streams is being reshaped. This includes machine learning, predictive analytics, and adaptive platforms. In the future, it could involve even more advanced systems that continue to redefine how work is done.

No matter the specific technologies, the core trend is that tasks once handled manually are shifting toward systems that can learn, adjust, and optimize on their own. IT leaders must rethink how teams are organized, how capabilities are built, and how the workforce is led through these changes.

# How Traditional Streams Are Being Redefined

- **Support and Operations**
  Teams that previously focused on resolving tickets and issues will increasingly spend time designing and overseeing systems that predict and prevent problems. Much of their work will involve supervising exceptions, tuning automated processes, and ensuring systems learn effectively.

- **Business Enhancements and Process Improvements**
  Instead of relying on periodic manual upgrades, these teams will integrate continuous optimization using intelligent tools that adapt workflows on the fly. This requires combining business process expertise with getting comfortable managing systems that make decisions in real time.

- **Analytics and Reporting**
  Business intelligence teams will move beyond building static dashboards. They will become stewards of data pipelines and ecosystems that feed automated processes, ensuring reliable, quality data for systems that act without manual oversight.

- **Strategic Projects**
  Many new initiatives will center on embedding intelligent capabilities. Whether it is adaptive customer interactions or automated risk detection, these projects demand teams skilled at experimentation and are adept at shaping outcomes alongside evolving systems.

- **Infrastructure and Security**
  Infrastructure teams will handle more than servers and networks. They will maintain platforms for large-scale data and real-time decision-making. Security will expand to protect systems that make choices on their own, including managing compliance and ethical safeguards.

## Consider a Dedicated Intelligent Systems Pathway

Some organizations take this even further by creating a specific pathway focused on intelligent automation and data-enabled business capabilities. This team takes clear ownership of:

- Identifying and prioritizing opportunities for advanced automation across the business.
- Designing, training, and continuously monitoring adaptive systems.
- Managing ethics, compliance, and regulatory requirements tied to autonomous processes.
- Measuring impact using continuous performance metrics, not just traditional ROI.

## What It Means for Your People and Leadership

As more work shifts to self-optimizing systems, the role of people changes. Engineers and developers move from building only static functions to shaping systems that learn and adapt. Data stewards and compliance professionals oversee how decisions are made by machines, ensuring trust and accountability. Leaders grow their ability to guide teams that are responsible for business outcomes even when some day-to-day decisions happen inside intelligent systems.

## What You Can Do Now

Preparing your IT organization for a future with intelligent automation focuses on progress versus perfection, taking steady steps forward rather than waiting for flawless plans. You can start taking practical steps right now.

- **Map where intelligent systems are already emerging**: Look across your current work streams. Identify where predictive tools, automated workflows, or systems that make decisions are already in place. This gives you a baseline to understand how your structures align.

- **Realign teams around business capabilities**: Move beyond technology silos and organize teams by the outcomes they enable, such as customer experience or supply chain responsiveness. As automation grows, these capabilities stay central.

- **Develop blended roles and cross-functional fluency**: Encourage staff to build skills that bridge operations, data, and business strategy. Support teams can learn about automation principles, while analysts see how their work fuels intelligent processes.

- **Make space for a dedicated pathway if needed**: If intelligent systems are becoming critical to your business, consider creating a team or center of excellence with clear ownership. This avoids fragmented responsibility and ensures consistent development and governance.

- **Invest in leadership education**: Provide training for leaders on how intelligent systems influence decision-making, how to guide outcome-based processes, how these changes impact staff, and how to navigate the ethics of automation effectively.

- **Keep your structures flexible**: Whatever pathways you set up, design them to adapt. As new forms of automation arrive, your teams should be able to pivot without getting trapped in rigid charts.

## The Takeaway

Get your IT organization ready for more autonomous, adaptive work by building people, structures, and leadership approaches that can evolve. This way, your teams are ready to help the business capture value from whatever intelligent systems come next.

# CHAPTER THIRTEEN

## Creating the Conditions for Innovation

### It Doesn't Start in a Lab – It Starts with Leadership

**TOO MANY ORGANIZATIONS** believe innovation begins by building a slick lab, launching a shiny pilot, or hiring a handful of outliers to play with emerging tech. In reality, sustained innovation begins with good leadership. It grows out of how you shape mindsets, set expectations, and design incentives across your entire organization.

If your culture relies on gatekeepers, rigid approvals, or is constrained by a fear of being wrong, no lab or pilot will fix it. Leading for innovation involves giving people room to explore, test, and learn without immediate penalties. It requires moving from protecting the status quo to inviting new thinking, even when it feels uncomfortable.

In one typical scenario from a regional bank, the CEO opened every quarterly leadership meeting by sharing something that did not go as planned and what was learned from it. This encouraged executives to discuss their own mistakes and course corrections. Over time, it normalized conversations about pilots, experiments, and lessons learned. The organization could see that curiosity and transparency were valued over perfection.

## Create the Safety and the Invitation to Try

Teams will not experiment if they believe one failed effort could stall their careers. Create a clear **permission to fail**, supported by a safety net that treats setbacks as learning, not personal flaws. People watch closely to see how mistakes are handled, and your leadership response sets the tone.

In a typical example from a mid-sized insurance organization, leadership encouraged business users to explore different workflows in a sandbox system. When early tests caused disruptions or revealed unexpected issues, these were treated as discoveries. The CIO reinforced that finding problems early was a success, not a mistake. Engagement shifted from cautious to curious.

Collaboration is equally critical. Rarely does innovation come from a single team. Breakthroughs usually emerge when people with different skills and experiences work together in a supportive environment to test ideas and learn.

For instance, in one retail setting, leaders overhauling customer loyalty programs invited not only marketing and IT, but also frontline store employees and customer support staff to co-design new ideas. This surfaced insights that would have been missed, leading to a simpler loyalty program and better tools for store teams.

This is how you move innovation from isolated sprints to a **continuous habit of improvement**. It happens when leaders intentionally build the right environment.

## A Firsthand Example: Innovation from the Field

I experienced this personally during my time working with a manufacturing and retailing company. Our top executives understood that real innovation would not come from inside headquarters. They encouraged us to get out into the world where our products actually lived. We spent time in our own stores and factories, visited dealers, and even invited retail customers to help rethink how our websites and customer tools should work.

One of the most powerful insights came from seeing our commercial division serve customers directly at job sites. We had mobile stores – trucks outfitted with hardwood floors, leather chairs, and air conditioning – that pulled right up to work sites. Customers were fitted for the right boots on the spot.

Being on site unlocked countless creative ideas. In addition to IT and product staff, we had cross-functional teams out in the field together, building camaraderie across departments. This led to practical innovations such as new inventory lookup tools, so store managers and field teams could ship products directly to customers if something were out of stock. We also developed automatic reorder capabilities for stores when inventory dropped to certain levels, along with many other improvements.

None of these came from a lab or a conference room. It happened because leadership made it a priority to get out of our comfort zones and into the real environments our customers lived and worked in.

# Quick Tips: Creating Conditions for Innovation

| Focus Area | Quick Tip |
|---|---|
| **Psychological Safety** | Make it safe to speak up. Reward candor; don't punish mistakes. |
| **Diverse Perspectives** | Bring in different voices. Innovation thrives on varied inputs. |
| **Curiosity and Exploration** | Encourage questions, experiments, and pilot projects. |
| **Clear Guardrails** | Set boundaries on risk and resources so people can explore wisely. |
| **Rapid Feedback Loops** | Test ideas quickly. Learn and adjust before scaling. |
| **Time and Space** | Protect time for thinking and creative work – don't let urgent tasks always win. |
| **Celebrate Small Wins** | Recognize progress, not just big launches. This builds momentum. |

## What You Can Do Now

- **Remove gatekeeping behavior**: Make it easier to try new approaches without layers of approvals or fear of being second-guessed.

- **Reinforce that smart failures are part of progress**: In a typical manufacturing scenario, leaders encouraged a small predictive maintenance pilot even though they expected early missteps. The first attempt did not predict much, but the team learned quickly. Within a year, plant downtime was reduced by 12 percent.

- **Set up safe spaces to experiment**: Whether through sandbox environments, rapid pilots, or cross-functional working groups, give teams clear places to test ideas.

- **Invite diverse voices**: Bring together people from across departments and roles. Often the best ideas come from unexpected perspectives.

- **Lead by example**: Be the first to question assumptions, to share lessons from ideas or strategies that did not work as planned, and to openly treat learning as a win.

## The Takeaway

Innovation starts with leadership that creates safety, invites collaboration, and celebrates thoughtful experiments, even if they initially fail. This builds a culture of continuous improvement that will serve your organization long after any single project.

# CHAPTER FOURTEEN

## What Great Teams Look Like

HIGH-PERFORMING, EMOTIONALLY MATURE teams are the result of leadership that sets clear outcomes, models accountability, and creates an environment in which people can do their best work.

This is why leadership is never a solo act. Your influence is ultimately measured by the teams you build and the culture they sustain. You can be the most strategic, technically savvy, or charismatic executive around, but if your teams cannot deliver together with trust and resilience, your impact will be shallow and short-lived.

### Hallmarks of Great Teams

Great teams are defined by how they show up for each other and the business every day. They reflect shared characteristics that reveal emotional maturity, healthy tension, and a deep commitment to outcomes over ego.

### Outcome-Focused

Anyone can check off a list. High-performing teams understand the broader outcomes they are driving. They ask how their work ties back to strategic objectives, how it serves customers, and how it moves the business forward.

They avoid the trap of simply doing what was assigned without challenging whether it still makes sense.

## Healthy Tension

Drama avoidance is a choice. Great teams lean into tough conversations, debate ideas vigorously, and challenge each other to get to better solutions. But they keep it grounded in respect and shared goals. They do not allow interpersonal friction to derail the work. This is where the Rules of Engagement come alive, reminding teams to stay fact-based, assume positive intent, and challenge with curiosity, not judgment.

## Accountability

Accountability reaches beyond meeting deadlines. In strong teams, people stay committed even when things get hard, they rebound quickly from setbacks, they take ownership for the full problem (not just their piece), and keep learning so they do not repeat the same mistakes. They see accountability as a source of pride, not a burden.

## Willingness to Pivot

Teams anchored in the status quo cannot innovate. High-performing teams balance confidence in their expertise with the humility to adjust when new data or opportunities emerge. They see changing course as a strength, not a weakness.

## Examples from Practice

- In many top-performing organizations, teams hold short retrospectives after milestones, not just to pat themselves on the back, but to explicitly identify what to carry forward and what to change. They normalize talking about where efforts went wrong without assigning blame.

- In client workshops, I have seen teams literally post their Rules of Engagement on the wall. This keeps them focused on outcomes, encourages direct feedback, and helps them call out drama or avoidance when it creeps in.
- In high-stakes projects, the best teams I've worked with did not hide problems. They surfaced risks early, held each other accountable for fixing them, and adjusted priorities together rather than retreating to protect their own turf.

## Quick Tips: What Great Teams Look Like

| Focus Area | Quick Tip |
|---|---|
| **Outcome-Focused** | Tie daily work to bigger goals. Ask how it serves customers and moves the business forward. |
| **Healthy Tension** | Debate ideas with respect. Lean into hard conversations without letting friction derail progress. |
| **Accountable** | Own the full problem, not just your piece. Stay committed through setbacks and keep learning. |
| **Willing to Pivot** | Be confident enough to adjust. Changing course when needed is a strength, not a weakness. |

| Use Rituals | Hold quick retrospectives to capture lessons. Post shared rules to guide team behaviors. |
|---|---|
| Surface Risks Early | Don't hide problems. Call out issues together and re-prioritize as a team. |

## What You Can Do Now

If you want your teams to look like this, start by modeling these behaviors yourself. Then make them explicit expectations for your leaders and teams.

- **Reinforce outcomes over tasks**: Keep bringing conversations back to why the work matters and how it ties to strategic goals.

- **Use your Rules of Engagement**: Post them, talk about them, and ask the team to hold each other to them.

- **Celebrate learning and resilience**: Recognize people who take smart risks, recover from setbacks, and stay committed under pressure.

- **Encourage constructive tension**: Make it safe to challenge assumptions and debate ideas without personal fallout.

- **Spot and reset drama**: When you see storytelling, blame, or avoidance start to creep in, pause the conversation. Re-center on facts and shared goals.

## The Takeaway

Your leadership legacy is defined by the quality of the teams you build and the culture they sustain as well as the systems and strategies you leave behind. When you see teams that focus on outcomes, embrace healthy tension, own their results, and keep learning together; that is what great looks like.

# PART IV – Operating in Complexity:

# Tactical Tools for Leading in a Fast-Moving, Risk-Aware World

IN THIS PART, we will step into the frontier where technology is evolving from tools of simple automation to agents capable of independent decision-making. We will explore AI's rapid progression from autonomy to agency, and how it is reshaping business operations, competition, and leadership.

This progression comes with profound responsibility. We will look at cybersecurity, ethics, data privacy, and risk, not just as safeguards, but as strategic enablers with guardrails that allow organizations to move faster and with greater confidence. You'll see how embedding these elements into a unified, business-led framework for IT ensures that innovation doesn't outpace accountability.

We will also tackle the reality that none of this happens in silos. Making AI work safely at scale requires intense collaboration across technology, business, compliance, and security teams. It involves aligning shared principles, openly addressing risks, and building the muscle to adapt as technology keeps accelerating.

You'll get a glimpse of what "healthy AI" truly looks like in practice, along with ways to measure your team's readiness across three critical domains. And we will close by offering a leadership manifesto: a call to action for guiding your organization through these uncharted waters with intention, integrity, and a deep commitment to shaping a future in which people and machines achieve more together.

# CHAPTER FIFTEEN

# From Autonomy to Agency

**AI IS MORE** than a single technology or even a single era. It is best understood as an unfolding arc, one that stretches from basic automation to something far more powerful: systems that can reason, decide, and act on their own. This journey, from autonomy to agency, shapes how organizations work, how leaders lead, and what it truly means to be ready for the future.

Understanding this evolution matters because it gives you a durable framework. The hype cycles will come and go. Specific tools will rise and fall. But these underlying shifts in what AI can do and how much it can be trusted to execute will define your decisions for years to come.

## Traditional AI: Automation Through Autonomy

The earliest and most familiar form of AI is not particularly intelligent at all. It is rigidly automated. These systems execute predefined rules on structured data. They are the invoice-matching tools that scan barcodes, the fraud detection engines that flag transactions based on thresholds, and the scripts that reconcile ledgers or trigger standard alerts.

They are excellent at doing the same task over and over with near-perfect consistency. They do not learn. They do not reason. They do not adapt to new situations. That makes them reliable, but it also makes them inflexible. They are best for basic, repetitive processes that rarely change.

## Generative AI: Augmentation Through Intelligence

Then came a more flexible, context-aware phase. Generative AI systems process unstructured data, draw connections, and produce new content. They write marketing copy, draft contracts, summarize research, generate code, and even help diagnose patient symptoms by analyzing complex case data.

These tools are smart, but they still lean heavily on human framing and oversight. They do not know which problems matter most. They do not initiate goals on their own. Instead, they act as supercharged co-pilots, accelerating work that once took days or weeks. They augment human capacity, making it easier to explore ideas, test concepts, and reach first drafts faster.

## Agentic AI: True Agency

Agentic AI systems suggest actions and then take them. They learn from outcomes, adapt strategies in real time, and orchestrate multistep workflows with minimal human intervention. A retailer might deploy an agentic system that dynamically adjusts pricing, updates supply chain orders and manages vendor communications without manual approvals. A property management company might let AI oversee the entire tenant lifecycle, from lease renewals to maintenance scheduling, only escalating when exceptions arise.

These systems bring extraordinary efficiency and speed, but they also raise the stakes. Once an AI can decide and act, it magnifies not just your opportunities, but also your risks. It blurs the lines of accountability in ways that demand new leadership guardrails.

## How This Evolution Fits Together

It is important to remember these are not neatly separated stages. Most organizations today use all three at once. They may still have rigid automation in payroll, experiment with generative tools for customer engagement, and cautiously test agentic platforms to run end-to-end operational processes.

What matters is clarity. You need to know what the business impact is of what you are automating, what you are augmenting, and what you are delegating. That clarity will drive your expectations, your risk tolerance, your governance structures, and ultimately, your readiness.

## A Timeless Lens

Figure 15.1 is a simple way to keep it all straight, no matter how technology evolves in the years ahead.

**Figure 15.1** The Evolution of AI:
From Autonomy to Agency

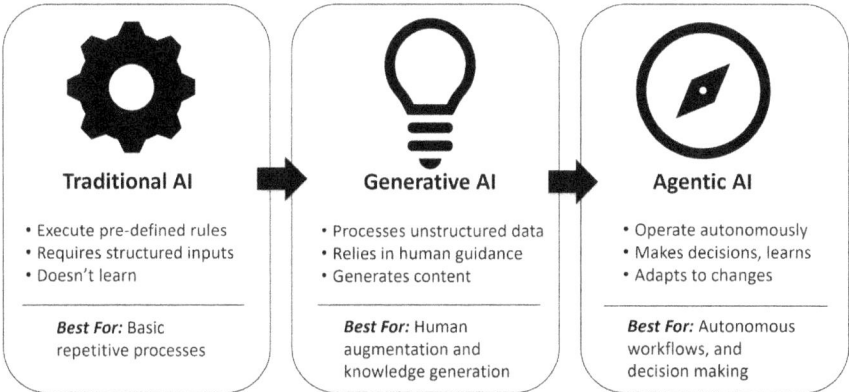

| Traditional AI | Generative AI | Agentic AI |
|---|---|---|
| • Execute pre-defined rules<br>• Requires structured inputs<br>• Doesn't learn | • Processes unstructured data<br>• Relies in human guidance<br>• Generates content | • Operate autonomously<br>• Makes decisions, learns<br>• Adapts to changes |
| *Best For:* Basic repetitive processes | *Best For:* Human augmentation and knowledge generation | *Best For:* Autonomous workflows, and decision making |

## Looking Ahead

The next chapter will help you assess whether your organization is truly ready for this future, not just technically, but strategically, ethically, and culturally. Before you decide how far to lean into agentic solutions or how aggressively to automate, you need a clear picture of your own maturity and risk appetite. Understand where your business is on this spectrum today, and where it wants to be tomorrow.

## What You Can Do Now

Actions and Quick Tips for Building Agency

- **Clarify Outcomes**: Be explicit about the bigger goals, not just the tasks. Help people connect their decisions to strategic impact.

- **Set Guardrails Together**: Establish shared principles for quality, risk, and decision-making. These keep freedom aligned with responsibility.

- **Build Skills and Judgment**: Invest in coaching, cross-training, and critical thinking so teams can navigate choices confidently.

- **Model Trust**: Show that you trust teams by letting them solve problems their way, then support them even if course corrections are needed.

- **Debrief and Learn**: After decisions, talk through what worked and what didn't. This grows agency by building collective wisdom.

# CHAPTER SIXTEEN

## Cyber, Risk, and Ethics as Strategic Enablers

### More Than an IT Concern

CYBERSECURITY, DATA PRIVACY, and digital ethics are no longer technical concerns to be handled by specialists alone. They have become business risks in every sense. They set on board agendas alongside financial, operational, and market risks. Study after study shows that boards and executive teams overwhelmingly see cyber as a core business issue, not just an IT concern.

Yet, many organizations still treat these areas narrowly. They see them as compliance hurdles or purely defensive measures. This mindset is outdated and dangerous. Trust has become a primary driver of customer loyalty, brand reputation, and even market valuation. Without clear leadership on security, privacy, and ethical technology use, companies undercut the very trust they need to grow.

### Guardrails That Enable, Not Block

The biggest misconception is that governance slows down development. In truth, clear guardrails speed up progress. They create clarity and safety, so teams know how to move without constant second-guessing.

Many people hear the word *governance* and think *bureaucracy*. In reality, good governance means clear responsibilities, lightweight controls, and shared owner-ship. It ensures decisions are made with eyes wide open.

Your enterprise transformation framework makes this clear. In today's world, many systems are selected directly by business units. Some are **advocated systems**, fully vetted and supported by IT. Others are **non-advocated systems**, chosen by business leaders without full IT over-sight. Either way, once a system is brought in, leaders own far more than vendor selection. They take on obliga-tions for security, privacy, integration, ongoing mainte-nance, and compliance.

This is where cyber, privacy, ethics, and architecture move from being obstacles to becoming strategic enablers. Leaders who understand these responsibilities from the beginning make better choices. They build trust with cus-tomers and reduce surprises later.

## The Gears of Enterprise Risk Collaboration

Governing cybersecurity, privacy, digital ethics, and data-driven decisions is a collaborative system with three inter-connected components that work like gears (see Figure 16.1). Each gear relies on the others to operate effectively. When aligned, they protect the business while enabling it to move with speed and confidence.

**Figure 16.1** The Gears of Enterprise Risk Collaboration

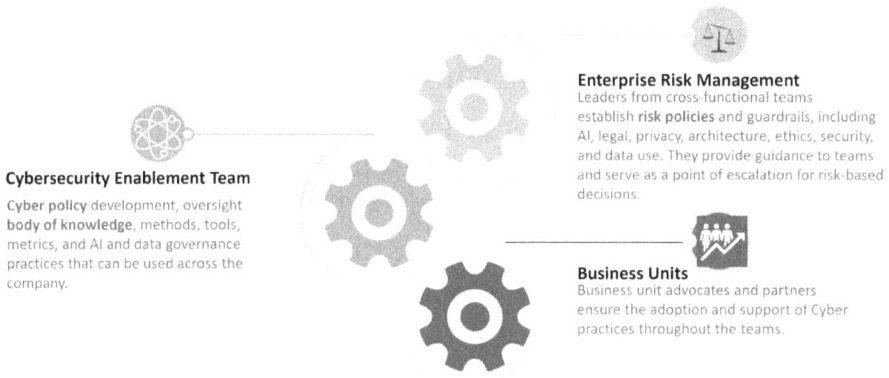

**Enterprise Risk Management**
Leaders from cross-functional teams establish **risk policies** and guardrails, including AI, legal, privacy, architecture, ethics, security, and data use. They provide guidance to teams and serve as a point of escalation for risk-based decisions.

**Cybersecurity Enablement Team**
**Cyber policy** development, oversight body of knowledge, methods, tools, metrics, and AI and data governance practices that can be used across the company.

**Business Units**
Business unit advocates and partners ensure the adoption and support of Cyber practices throughout the teams.

## Risk Enablement Teams

These are your domain experts. They develop policies, oversee standards, and maintain the body of knowledge that defines how the organization approaches legal, privacy, ethics, cyber, and responsible technology practices.

They create the methods, tools, and metrics that all teams across the company can use throughout the decision lifecycle. They also provide ongoing support to help business and IT leaders apply these principles in real-world contexts. Their role goes beyond issuing blanket mandates. They build clear, practical frameworks that help the organization make smart, risk-aware choices.

## Cross-Functional Risk Leaders

This group includes leaders from across the business who come together to shape risk policies, set guardrails, and resolve decisions that span departments. These leaders bring experience from operations, product development, marketing, IT, legal, and customer experience. Their diversity ensures that risk decisions are grounded in how the business actually runs.

They also serve as escalation points when decisions reach beyond one team. This is where tradeoffs are discussed, such as balancing speed to market with customer data protection or aligning automation efforts with fairness and transparency principles. This group helps maintain balance between protecting the business and helping it grow.

## Business Units and Advocates

This is where the day-to-day work happens. Business units operate with autonomy but within a structure. They adopt and follow the cybersecurity, privacy, and ethical practices developed by domain experts and reinforced by enterprise leaders.

Within these units, advocates and team leads are critical. They bring practices to life in projects and workflows. They translate policies into local action, identify issues early, and escalate risks when needed. Just as importantly, they provide feedback to improve the system. This keeps governance connected to operational reality. When business units are supported rather than micromanaged, they become essential allies in protecting the company and building customer trust.

## Why This Model Works

These three gears form a connected system. Risk knowledge flows from domain experts. Priorities and decisions are shaped by cross-functional leaders. Business units put the principles into action, and their insights help refine the model.

This collaborative approach turns cybersecurity, privacy, and ethics into innovation of enablers. It replaces reactive compliance with a culture of shared ownership. And it gives the business the clarity and confidence it needs to move forward without compromising what matters most.

## A Unified Model for Business-Led Risk Management

Your risk governance framework shows how this all fits together. It illustrates four interconnected domains that keep business-led technology initiatives safe and scalable:

- **Legal requirements:** meeting contractual, regulatory, and statutory obligations.
- **Cybersecurity:** protecting company and customer data, preventing disruptions and reputational harm.
- **Privacy:** honoring promises to customers and regulators on data collection, use, and storage.
- **Solution architecture:** making sure new systems integrate properly and do not become fragile one-offs.

Risk is not static. It evolves across a lifecycle that starts by identifying threats and obligations, continues by putting controls in place, and then requires ongoing monitoring and adaptation. It only ends when a system is decommissioned and data is retired responsibly. These guardrails work together like gears, each reinforcing the other. This is how you enable business-led innovation while keeping customer trust and long-term sustainability intact.

## Trust and Ethics as Strategic Assets

Many companies still think about ethics and compliance narrowly. They focus on avoiding fines or regulatory action. That is important, but it is a limited view.

Trust is a strategic asset. Customers are far more likely to share data, embrace new services, and deepen relationships when they believe your company uses technology responsibly. Practices such as fairness, transparency, and privacy by design build long-term loyalty.

This is why digital ethics cannot be delegated only to IT or risk teams. It is a leadership issue. Organizations that prioritize responsible technology create a foundation that supports continuous innovation. They also avoid the kinds of scandals and reputational damage that can erase years of investment overnight.

## Exercising Risk Preparedness

Good governance is tested and refined through practices that show how your organization responds under pressure. This is where Business Impact Assessments (BIAs) and tabletop exercises come in.

Reaching beyond a catalogue of systems and downtime thresholds, a BIA connects technology risks to business realities by asking what matters most. Which processes protect your biggest revenue streams? Which ones ensure customer trust and compliance? A BIA puts real numbers on these questions, so leaders understand exactly what an hour of downtime costs. It also sets meaningful Recovery Time Objectives (RTO) and Recovery Point Objectives (RPO). That turns risk-funding decisions into business-led discussions.

Tabletop exercises take these priorities and stress-test them. They bring IT, operations, legal, HR, and executives together to work through realistic crisis scenarios. This is where hidden dependencies and gaps in decision paths come to light.

Practicing a ransomware scenario, for example, goes well beyond the technical response. It forces clarity on who can approve payments, how customers and regulators are notified, and what happens if data restoration takes longer than planned. These exercises keep governance alive and practical. When done consistently, they turn cybersecurity, data protection, and ethics from abstract principles into living capabilities. They protect customer trust and give leaders the confidence to keep pushing forward.

## What You Can Do Now

- **Use a BIA to set clear priorities**: Identify critical business processes, quantify the impacts of downtime, and establish realistic RTO and RPO targets. Tie resilience investments directly to these priorities.

- **Run tabletop exercises with business leaders at the table**: Move beyond IT-only drills. Bring in legal, HR, operations, and executive decision-makers. Use real scenarios to practice making hard calls under stress.

- **Embed guardrails across the lifecycle**: Make sure security, privacy, compliance, and architecture are addressed from requirements and design through procurement, operations, and system retirement.

- **Establish principles for ethical tech use**: Set clear expectations for fairness, transparency, and privacy by design. Make these part of how new initiatives are approved and measured.

- **Reinforce trust as a source of growth**: Remind teams that protecting customer data and using technology responsibly does more than avoid penalties. It builds deeper relationships that lead to more opportunities.

## Quick Checklists to Guide Your Teams

### *Business Impact Assessment Checklist*

- ✓ Identify critical business processes.
- ✓ Quantify impacts over time for financial, customer, and compliance exposure.
- ✓ Set realistic RTO and RPO targets.
- ✓ Map key systems, vendors, and dependencies.
- ✓ Review after major business changes or at least once a year.

### *Tabletop Exercise Checklist*

- ✓ Include leaders beyond IT such as legal, HR, operations, communications, and executives.
- ✓ Use realistic scenarios such as ransomware or major outages.
- ✓ Test escalation paths and decision authority.
- ✓ Validate how external communications will work for customers and regulators.
- ✓ Capture lessons learned and update playbooks.

### *DR and Cyber Playbook Readiness*

- ✓ Confirm documented recovery steps are current and tested.
- ✓ Validate contacts for key vendors and external support.
- ✓ Make sure financial approval levels for emergency spending are clear.
- ✓ Keep offline copies of playbooks and contact lists.
- ✓ Schedule failover or recovery tests regularly.

## The Takeaway

Cyber, privacy, and ethics are the foundations that makes innovation possible at scale and with confidence. When treated as strategic enablers, they allow your organization to move faster and earn trust that compounds over time.

# CHAPTER SEVENTEEN

## Are You Ready? Building Tech Readiness in the Age of AI

### More Than Technical Preparedness

AI **IS NOT** coming. It is already here. Its influence on jobs, workflows, and entire industries grows every day. As a leader, you do more than implement technology. You help people think differently, keep learning, and lead with values that unite trust and innovation.

AI is most powerful when it is guided by human priorities. It should elevate the quality of work, help teams solve bigger problems, and create more space for creativity and meaningful connection. This only happens if your organization is truly ready. In this chapter, we'll reinforce that readiness is not only technical. It is deeply behavioral and cultural, building the mindsets and habits that make your organization adaptable, resilient, and always learning.

### The New Leadership Edge: A Behavioral Readiness Framework

After decades of helping companies prepare for major technology shifts, I see three broad behavioral themes that consistently stand out. They are habits and cultural norms that become a real competitive edge. The following are the three attributes that separate organizations and leaders who thrive with AI from those who struggle to keep up:

## 1. **Mindset and Adaptability**

This section will highlight how people approach change, uncertainty, and new ways of working:

- **Curiosity:** Seeking out new knowledge, tools, and perspectives. You see it in people who ask thoughtful questions, explore new AI tools, and look outside the industry for ideas.
- **Learning agility:** Quickly learning, unlearning, and applying lessons in new situations. These people adapt to new systems and carry insights from one project to another.
- **Resilience:** Staying effective under ambiguity or pressure. They reframe setbacks and keep moving forward during change.
- **AI openness:** Willingly experimenting with AI and embracing it as an enabler, not a threat. They try new tools and encourage peers to use them.
- **Change advocacy:** Supporting and championing organizational transformation. Leaders help colleagues adapt and communicate changes constructively.

Organizations measure this by tracking learning hours, time to proficiency, how people handle setbacks, and participation in change efforts. Peer and manager feedback gives the clearest signal of who is embracing new ways of working.

## 2. **Execution and Influence**

This focuses on how people drive work forward and engage others:

- **Initiative:** Acting without waiting to be told, proposing improvements, and building small pilots.

- **Accountability:** Owning outcomes, meeting commitments, admitting mistakes, and following through.

- **Collaboration:** Working effectively toward shared goals, building on others' ideas, resolving tension, and sharing credit.

- **Influence:** Shaping decisions through credibility and communication, championing new tools, explaining ideas clearly, and securing buy-in.

- **Value orientation:** Prioritizing work that drives business and customer outcomes.

Organizations look for these through innovation logs, project activity, feedback processes, and by seeing who consistently moves efforts forward tied to what truly matters.

## 3. **Enablement and Systems Thinking**

The following themes connect dots, scale knowledge, and look beyond silos:

- **Digital fluency:** Using digital and AI tools confidently, automating tasks, and integrating workflows.

- **Knowledge sharing:** Documenting solutions, mentoring others, and leading demos that lift the entire organization.

- **Ethical empathy:** Considering the human impacts of decisions, watching for bias, promoting fairness, and protecting trust.

- **Systems thinking:** Understanding how parts connect across the business, designing scalable, integrated solutions.

This shows up in how people use tools, contribute to shared resources, coach peers, test ethical scenarios, and get feedback from cross-team work. These behaviors ensure AI does not live in a silo, but become a capability the whole organization can use responsibly.

## AI Readiness at a Glance

| Attribute | Definition | How It Shows Up | How to Measure |
|---|---|---|---|
| Mindset and Adaptability | | | |
| Curiosity | Seeks out new knowledge, tools, and perspectives | Asks thoughtful questions, explores new AI tools, learns across industries | Learning hours, peer or manager feedback |
| Learning Agility | Quickly learns, unlearns, and applies knowledge | Adapts to new tools, applies lessons in new contexts | Time to proficiency, success in project rotations |
| Resilience | Stays effective under ambiguity or pressure | Reframes setbacks, keeps momentum during change | Bounce-back time, stress assessments |

| | | | |
|---|---|---|---|
| AI Openness | Willingly experiments with AI, sees it as an enabler | Tries new tools, encourages peers to use AI | Usage logs, enablement participation, sentiment surveys |
| Change Advocacy | Champions organizational transformation | Helps peers adapt, communicates changes constructively | Participation in change efforts, peer feedback |
| **Execution and Influence** | | | |
| Initiative | Acts without prompting, solves problems proactively | Proposes improvements, builds MVPs, leads pilots | Innovation logs, project activity |
| Accountability | Owns outcomes and delivers consistently | Meets commitments, owns mistakes, follows through | Goal attainment, 360 feedback |
| Collaboration | Works toward shared goals, manages tension well | Builds on others' ideas, shares credit | Peer NPS, team retrospectives |
| Influence | Shapes decisions through communication and credibility | Champions new tools, gains buy-in | Stakeholder feedback, idea adoption rates |
| Value Orientation | Focuses on work tied to business and customer outcomes | Aligns efforts to impact, prioritizes high-value activities | Manager ratings, contribution to KPIs |
| **Enablement and System Thinking** | | | |
| Digital Fluency | Uses digital and AI tools confidently | Automates tasks, integrates workflows | Tool usage data, fluency assessments |
| Knowledge Sharing | Shares learning openly to uplift the organization | Documents solutions, mentors, leads demos | Wiki contributions, peer coaching hours |

| | | | |
|---|---|---|---|
| Ethical Empathy | Considers ethical and human impacts of decisions | Flags bias, promotes fairness, builds trust | Scenario testing, ethics training, trust ratings |
| Systems Thinking | Understands how parts connect in broader the organization | Designs scalable, integrated solutions | Feedback from cross-team work, holistic solution design |

# What Healthy AI Looks Like

Readiness includes knowing what healthy, responsible AI looks like, beyond quick pilots or flashy tools. Healthy AI is built on solid foundations and guided by leaders who see it as a strategic capability.

Five key elements stand out:

1. **Access to rich, trusted data.**
   Strong AI is built on well-governed data across the organization, including structured transactions and unstructured content such as documents and service chats. This foundation turns AI from guesswork into real intelligence.

2. **Human judgment from domain experts.**
   AI works best when paired with business insight. Experts test assumptions, validate outputs, and keep solutions grounded.

3. **Solution architecture with people in control.**
   Skilled architects make sure systems can evolve while letting humans adjust course. This builds trust and avoids risky black-box decisions.

4. **Scalable multi-agent ecosystems.**
Many organizations are moving to environments in which multiple intelligent agents work together. Designing for this defines clear roles, creates shared data platforms, and sets oversight in place.

5. **Cybersecurity and risk governance.**
AI brings new risks as well as opportunities. Responsible programs bake in privacy, zero trust, accountability, and model checks from the start. Far from slowing innovation, this gives teams the confidence to accelerate and take smarter risks.

When these foundations are in place, AI sparks creativity. It solves bigger problems and allows the business to adapt quickly. It becomes a strategic asset built on trust.

## What You Can Do Now

- **Keep humans at the center:** AI readiness puts people first. It prepares humans to lead, question, guide, and adapt as intelligent systems take on more work. Make sure every initiative looks at how AI changes your teams' daily experience and helps them focus on meaningful work.

- **Start with conversations, not just tools:** Talk with teams about how AI may change their work, where it can remove low-value tasks, and how it might raise customer expectations. Reduce fear by treating this as a shared learning adventure.

- **Spot and reinforce the right behaviors:** Look for examples of innovative learning, ethical empathy, and systems thinking. Recognize these out loud. This signals what the organization values.

- **Pilot responsibly:** When trying new AI solutions, pair them with domain experts to keep work grounded. Build small successes that teach everyone how to scale with care.

- **Embed fairness, privacy, and risk checks:** Even early experiments should consider how decisions impact customers and communities. This builds trust from the start.

- **Measure and share progress:** Track learning hours, tool adoption, sharing across departments, and how often teams pivot based on new insights. Make these visible so the culture evolves together.

## The Takeaway

AI readiness goes beyond systems or data. It truly is about people. Humans are leading, questioning, and guiding technology so it serves what matters most. AI should always be for humanity, led by people who consider its impact on teams, customers, and communities. Build these habits, mindsets, and shared responsibilities now, and your organization will be ready for whatever technological revolution comes next.

# CHAPTER EIGHTEEN

## Leading What Machines Can't: Thriving in the Age of AI

**WE ARE LIVING** through one of the most significant shifts in the history of work. Generative AI and agentic solutions, systems that not only analyze and predict but also decide and act, are reshaping how organizations operate. It is tempting to see this as just another technology wave: smarter models, faster automations, new dashboards. But the real story is about what it means to lead and create value in a world where machines can now do much of what once required people.

AI certainly speeds up tasks, and today it is also reshaping them. Hours of manual work are now reduced to minutes. Insights appear before humans even think to ask. Autonomous decisions flow through marketing campaigns, supply chains, financial planning, and customer service. This does place some roles at risk. Many activities will be automated, certain jobs will be redefined, and layers of traditional oversight may shrink.

At the same time, what humans uniquely bring – judgment, empathy, influence, creativity, and a sense of purpose – becomes even more important. The future belongs to leaders who know how to blend people and intelligent systems to unlock outcomes neither could achieve alone.

## How Leadership Is Already Changing

Leadership is already evolving under the pressure and promise of these new tools. We are seeing:

- Less time on transactional oversight. AI handles monitoring, tracking, and summarizing.
- More emphasis on shaping vision, strategy, culture, and ethical guardrails.
- New accountability. Leaders are now responsible not just for people, but for the systems and AI agents they put in place.

The most critical leadership strengths are the most human ones. Leaders must craft purpose, build trust, navigate ambiguity, influence others, and ensure technology is used responsibly.

## A Future Leadership Manifesto

This lifts the bar beyond personal philosophy into a set of promises that can help anyone lead with purpose and confidence in an AI-driven world.

### *Future Leadership Manifesto: Leading What Machines Can't*

- **Clarity of Purpose:** We will lead with clarity, not just efficiency. Machines optimize. Humans inspire.

- **Humanity:** We will stay deeply human. We will invest in trust, empathy, and ethical decision-making.

- **Partnership:** We will partner with AI, not abdicate to it. We will use it to amplify human potential, never to dodge responsibility.

- **Curiosity and Learning:** We will foster curiosity and learning. We will explore how technology can transform value while staying vigilant to its risks.

- **Accountability:** We will own the systems we deploy. AI does not absolve our accountability.

- **Adaptive Culture:** We will shape cultures that adapt and thrive. Technology moves fast. Our organizations must keep pace.

## How to Stay Indispensable as AI Grows

So what does this mean practically? Here are the habits, mindsets, and skills that will keep you relevant and indispensable.

- **Get AI fluent:** Understand what these systems can and cannot do. Learn their biases, data needs, and blind spots.

- **Elevate your human edge:** Develop deeper empathy, influence, creative thinking, and sound judgment.

- **Become a master problem-framer:** AI solves tasks. Humans decide which problems are worth solving. That is where your highest value lies.

- **Build strong, diverse relationships:** Trust and collaboration will matter even more as technology changes how people work together.

- **Lead with transparency and ethics:** Make sure your AI initiatives are explainable, fair, accountable, and maintain trust.

- **Stay curious and adaptable:** The half-life of skills continues to shrink. Keep learning, unlearning, and relearning.

- **Focus on outcomes, not outputs:** Let machines handle transactions while you drive strategic impact.

## What You Can Do Now

This may all sound big and abstract, but you do not have to wait for a major corporate initiative to start. You can begin today. Here are some small, meaningful actions to help you and your team prepare.

- **Try an AI tool on a real work task:** Pick something low-risk, such as summarizing a long document or drafting a first-pass analysis. Reflect on how it changes the work.

- **Ask your team members where they see opportunities:** Invite them to identify routine work that could be augmented by AI.

- **Host a discussion on AI ethics:** Use real scenarios from your business. Discuss where you would draw lines.

- **Map out a simple "AI plus human" workflow:** Take a process and sketch what changes if you add an AI step.

- **Read one new article or guide each week:** Build your comfort with the language and possibilities.

- **Talk openly about fears and expectations:** Help your team see that curiosity and caution can coexist.

These are the small moves that build future-ready muscles. The more you practice, the more natural it becomes to shape work where people and AI complement each other.

## Closing Thought

The age of GenAI and agentic solutions will rewrite business playbooks. The leaders who thrive will not be the ones who resist change or rush in without thought. They will be the ones who use these new tools to elevate what only humans can offer: purpose, trust, creativity, and the resilience to adapt.

# PART V – Leading for the Long Game:

## The Internal Work of Being a Trusted, Lasting Presence

IN THIS FINAL part, we will come back to where real change always begins and ends: with people. As technology accelerates and the future grows less certain, it is our humanity that becomes the ultimate differentiator. Here, we will focus on how to lead with grit, calm, and steadiness, especially when the pressure rises. Leading this way keeps you grounded, builds trust, and engages in the hard conversations that strengthen teams and build resilience.

We will explore how to prepare our people for what is next, not just through new skills, but by creating an environment in which they feel supported, understood, and part of the journey. We will look at coaching as fuel for the long game, one of the most practical ways to keep growing yourself so you can better grow others. We will dive into self-reflection as the secret weapon of personal growth, the practice that sharpens your judgment, tempers your reactions, and deepens your impact.

And we will close by exploring legacy and trust, the imprint of your leadership on others, measured not by the systems you build alone, but by how people experience your presence, your consistency, and your care. In the end, this wave of change is about people. Your role as a leader is to guide them through uncertainty with clarity and compassion, so they are stronger and so are you.

# CHAPTER NINETEEN

## Coaching: Fuel for the Long Game

**WHEN I SIT** down with leaders, I often ask how they're investing in their own growth. Not the growth of their teams or organizations, but their personal growth. This is where many pause. They've been so focused on delivering for everyone else that they've forgotten to sharpen their own edge.

It's always struck me as curious. We coach our kids. Athletes at every level have coaches who help them see what they can't, stretch their abilities, and stay accountable. Yet, far too few business leaders give themselves that same support. Why do we treat ourselves differently? Why do we expect to navigate increasingly complex challenges without the kind of guidance we readily provide for others?

Coaching is one of the best ways I know to change that. It is how you stay ready for what comes next. The world keeps moving. Technology evolves, markets shift, and organizations adapt. Through all of it, leaders who keep learning and growing are the ones who last.

Coaching gives you a dedicated space to step back, gain perspective, and build new capacity for whatever lies ahead. It sharpens your thinking, strengthens your resilience, and prepares you to guide others with clarity and confidence. Leaders who keep learning and growing are the ones who thrive.

# Choosing the Right Coach

Finding the right coach starts by getting clear on what you need. What are you hoping to accomplish? Where do you feel stuck or under pressure? How will you know if the time and energy you put into coaching was worth it?

Then think about style and experience. Do you want someone who's carried the same weight you do – who has been a CIO, led big teams, or sat at the executive table? Or would you rather work with someone who brings a completely different perspective and challenges how you think?

Pay close attention to trust and chemistry. The best coaching relationships work because you can be open and honest. You need to feel safe enough to explore your blind spots without getting defensive. That takes the right fit.

## A Framework for Selecting Your Coach

Choosing a coach is too important to leave to chance. Use this simple framework to clarify what you're looking for. The clearer you are, the better your fit will be and the more powerful your coaching experience becomes.

| Area to explore | Questions to consider |
|---|---|
| **Your goals** | What are you hoping to achieve, both short and long term? What specific challenges are you facing? |
| **Success measures** | How will you know the coaching was worth it? What changes or outcomes would show real progress? |

| | |
|---|---|
| **Industry and role experience** | How important is it that your coach has walked in your shoes, whether as a CIO, business leader, or in your field? |
| **Coaching style** | What leadership style would best serve your team right now – someone decisive, someone who nurtures and supports, or someone who brings people together to solve problems? What approach pushes and supports you best? |
| **Use of tools and frameworks** | Do you value practical tools and structured methods, or prefer more open conversation? |
| **Values and chemistry** | What personal qualities matter most – integrity, empathy, accountability? How important is trust and comfort to you? |
| **Communication preferences** | How do you want to meet (video, phone, in person), how often, and what kind of follow-up or support do you expect? |

Take your time reflecting on these questions. The best coaching relationships start with a clear understanding of what you want, how you like to work, and what success looks like to you.

## Preparing for the Work

Coaching is a commitment. It is not a series of polite conversations. It is work that can feel uncomfortable at times because growth often does. Start by asking yourself honest questions:

- What do you want most in your career or life?
- Is something important missing right now?
- Where are you settling for good enough instead of pushing for great?

- If you knew you couldn't fail, what would you go after?
- How willing are you to hear hard feedback and act on it?

The more candid you are about your hopes, frustrations, fears, and patterns, the more powerful coaching becomes. You're the one driving the outcomes. A coach is there to challenge, support, and hold up the mirror so you can see yourself more clearly.

## Guided Versus Willing: How Mindset Shapes Your Growth

Not everyone comes to coaching the same way. Some leaders start because their boss, HR, or a board member suggested it. Others seek it out themselves because they understand that growth takes real work. Here's how these two mindsets usually show up.

|  | Guided or Referred | Willing and Engaged |
|---|---|---|
| **Reason for coaching** | Doing it because someone suggested or mandated it | Actively seeking growth and fresh perspective |
| **Openness** | May be cautious or defensive at first | Curious, honest, and ready to stretch |
| **Focus** | Often trying to prove they're already good enough | Exploring blind spots and untapped potential |
| **Energy** | Passive participation, waiting to be led | Drives the agenda, takes responsibility |
| **Outcomes** | Usually incremental changes | Often experiences significant transformation |

The biggest difference I've seen over decades of coaching isn't talent or experience. It's willingness. Be the kind of leader who chooses to grow, not one who has to be convinced. That choice will change everything for you, your team, and the legacy you leave.

## Be the Kind of Leader Who Coaches Too

Coaching doesn't just help you. It changes how you show up for your team. You start asking better questions. You stop jumping straight to answers and solutions. You become more present, more tuned in, and more committed to helping others grow.

That is how strong cultures are built. Not by pushing people to deliver at all costs, but by leading them through challenges in ways that build confidence, skill, and resilience.

When you invest in being coached, you model that growth is not optional. It is how leaders stay ready for tomorrow.

## What You Can Do Now

- **Get clear on your goals:** Write down what you want most right now. Identify where you feel stuck or unsatisfied.

- **Reflect on your patterns:** Notice where you tend to play it safe or hold back. Think about what might change if you stepped forward differently.

- **Explore coaching options:** Look for a coach whose experience, approach, and style fit what you need. Trust your instincts about chemistry and comfort.

- **Make a real commitment:** Be ready to do the work. Growth takes honesty, effort, and openness to discomfort.

- **Practice coaching your team:** Ask more questions. Hold space for people to think. Support their growth, not just their delivery.

# CHAPTER TWENTY

## Leading with Humanity and Grit

### What Grit Really Looks Like

**GRIT IS OFTEN** misunderstood as merely a constant hustle or pushing yourself and your teams to exhaustion. True grit is quieter and more lasting. It manifests as emotional presence, calm under pressure, and with consistency that builds trust over time. Be the person your team can rely on, especially when things get tough.

This kind of leadership avoids creating urgency where none is needed and doesn't drive teams with fear or drama. Instead, it creates an environment in which people feel respected, heard, and valued. This is how they become stronger and better equipped to handle whatever comes next.

### Leading with Humanity Means Preparing People, Not Shielding Them

Leading with humanity is sometimes confused with protecting people from hard truths. That is not what it means. It sees people as individuals, recognizes their ambitions and concerns, and equips them to become more resilient. Stay aware of how change impacts people since every new strategy, system, or process ultimately becomes someone's responsibility. They are the ones who carry it forward.

Leadership centers on people. It builds the kind of environment in which they know you have their back. You trust your gut when you see potential in someone, even when it is not obvious on paper. You guide people through uncertainty so they come out stronger on the other side.

## Everything We Do Comes Back to People

It is easy to get caught up in technology, budgets, and timelines. But none of that works without people. Real people bring plans to life. They want to be led by someone who sees them, respects them, and believes in what they can become. They want to be challenged but also supported. They want to feel that their work matters.

When teams know you care about them as people, not just as resources, they give more of themselves. They stay engaged even during difficult transitions. They adapt more quickly, and they bounce back faster after setbacks. That is how grit becomes a trait of the entire team, not just an individual quality.

## Staying Calm and Consistent Under Pressure

Grit shows up most clearly in how you handle stressful moments. Leaders who stay grounded help their teams stay grounded. They keep meetings focused on facts and next steps instead of panicking. They give people room to solve problems without adding more tension.

Don't ignore problems or hide your own stress. Show how to work through issues constructively. Your calm becomes an anchor for others. Your consistency provides stability, even when the future is not fully clear.

# Building Resilience Through Trust and Honest Conversation

Some of the most meaningful experiences I have had as a leader did not come from big projects or impressive metrics. Some of my most satisfying wins resulted from taking chances on people and watching them grow.

At one company, I hired the barista from our small on-site café. She was curious, always asking what people did, and kept track of coffee orders by time of day just to understand the patterns. I learned she had a master's degree in music and was looking for steady work, but her chosen field was hard to break into, especially in our small town. I offered her an entry-level job as a SharePoint administrator. She accepted, we trained her, and she never looked back. She became a standout contributor, helped drive major initiatives, and is still thriving in her IT career today.

Another time, I hired someone who had been in IT for years but had stepped away to raise children. When she was ready to return, I trusted my instinct even though others were unsure. She quickly proved her value and has since grown into a senior role, still making a strong impact at the same company.

And then there was the team member who saw how AI was about to change our organization. He came to me with a proposal that might have sounded risky. He had analyzed his own job and laid out exactly how it could be automated. He was not trying to protect his role. He was using facts and logic to show how the company could work more efficiently, even though it would make his own position unnecessary. What stood out most was what he said after. "I learned the skills that landed me this job. I can do it again. I just need the chance." Together, we moved forward with the plan he helped design. His job was eliminated, just as he predicted. But he did not stop there. He learned new skills, moved into a different part of the business, and started making an impact in a completely new way.

Rather than painting over hard realities, he faced them directly and had the courage to build something new. As a leader, my role was not to shield him from change. I supported his willingness to grow and gave him the space to prove what he could do next.

## What You Can Do Now

- **Stay present and visible:** Be there for your team, not just during crises but as a steady part of everyday work.

- **Watch your reactions:** Your calm helps others hold steady under pressure.

- **Ask questions that show respect for people's thinking:** Invite them to shape solutions with you.

- **Keep your routines:** Maintain one-on-ones and team check-ins even during intense projects.

- **Recognize steady effort:** Praise resilience and thoughtful contributions, not just last-minute saves.

- **Use setbacks as shared learning moments:** Help your team process mistakes together so they grow stronger, not more cautious or fearful.

- **Trust your instincts on people:** When you see potential, take a chance. Support them, train them, and give them opportunities to stretch.

## The Takeaway

Leading with grit shows up as the person others know they can rely on. It combines emotional presence, calm under pressure, and respect for people that helps them become more resilient. This drives lasting performance. It is also why people will continue to follow you with whatever comes next.

# CHAPTER TWENTY-ONE

## The Power of Self-Reflection

### The Secret Weapon of Leadership

**SELF-REFLECTION IS THE** secret weapon of personal and professional growth. The most effective leaders build space into their lives to examine their choices, reactions, patterns, and blind spots. They make time to think, assess where they are strong, and recognize where they need to grow. This separates good leaders from the truly great ones.

Leaders who engage in honest self-reflection become more authentic, resilient, and clear-headed. They sharpen their judgment, deepen their humility, and build the kind of credibility that draws others in. They also recover more quickly from setbacks because they take time to learn, adjust, and try again. Everything in this book points to one truth. If you want to lead teams, navigate uncertainty, and build organizations that thrive, you must first lead yourself. Self-reflection is how you start.

### Why It Matters

Self-reflection pays off in many ways.

- **Awareness.** You cannot build up your strengths or improve on your weaknesses if you do not see them clearly.

- **Growth.** You only get better at that which you are willing to examine honestly.

- **Authenticity.** Knowing yourself helps others trust you more because you show up as real and grounded.

- **Resilience.** Reflection helps you learn from mistakes so they do not become repeated failures.

- **Better decisions.** When you understand your own values, triggers, and patterns, your choices become clearer and more aligned.

## 10 Practices to Strengthen Self-Reflection

Here are 10 simple practices you can start using right now. Each one helps build the habit of slowing down long enough to learn from your own experiences.

1. **Make it a habit, not an event.** Growth happens in small steps over time. Build dedicated moments of reflection into your routine, such as a standing meeting with yourself.

2. **Keep a leadership journal.** Writing down what you are learning forces clarity. It captures lessons and patterns that might otherwise slip by. Over time, your own notes become a map of your growth.

3. **Look in the mirror, not just out the window.** It is easy to blame circumstances or other people. True growth starts when you ask what your role was in any outcome. Be willing to see how your own decisions, reactions, or blind spots played a part.

4. **Seek honest feedback.** Invite people you trust to tell you where you might be missing something. Make it safe for them to be candid. Feedback is the fastest route to self-awareness.

5. **Practice asking yourself better questions.** Move beyond "Why did this happen to me?" to "What can I learn from this?" or "How could I approach this differently next time?" Keep in mind that the quality of your questions often determines the quality of your insights.

6. **Slow your reactions.** When something triggers you, pause before you respond. Notice your first emotional reaction. Ask yourself what is driving it. Pausing in the moment between stimulus and response creates space for growth.

7. **Review key interactions.** At the end of a day or week, replay important conversations in your mind. What went well? Where did your emotions steer you off track? What would you repeat, and what would you change?

8. **Watch for patterns.** Over time, your reflections will reveal themes. Maybe you shut down in conflict or avoid certain types of decisions. Maybe you thrive in complex problems but get impatient in slow-moving situations. Knowing your patterns is the first step to intentionally shaping them.

9. **Keep your values front and center.** Reflect on whether your actions align with your values. The most respected leaders are consistent. They make decisions that match what they say is important.

10. **Remember to be kind to yourself.** Focus on honesty and growth. Treat yourself the same way you would treat a valued team member who is learning and evolving.

## A Story That Changed My Leadership Forever

One of my own most powerful lessons in self-reflection came early in my career. I was the CIO when my boss challenged me to learn more about the business. He already trusted that I understood technology, so he gave me responsibility for a team outside the IT sphere. It was a well-run group, led by one of the best leaders I have worked with in my career.

Even though this leader reported to me, she brought me under her wing and taught me the business. One day, she came into my office just as I was finishing a call with another colleague. That person had been complimentary of my leadership. After I hung up, I turned to her and asked, probably a little too proudly, if she thought I was a good leader.

She paused and then said something that took me by surprise. "You could be."

I was stunned, maybe a little hurt, but I had the presence of mind to listen. She told me there was one thing holding me back. I did not give honest feedback. I was too nice. She gave me specific examples. We finished our meeting, but her words stayed with me all day.

The truth was, she was exactly right. I am a peacemaker by nature. I avoid confrontation without even realizing it. Her honesty, and my willingness to look inward, changed everything for me. I realized that by holding back feedback, I was not helping anyone. It limited their growth and mine. If people never hear what they need to improve, how can they get better?

We talked more in the weeks that followed. She helped me see that giving direct feedback, even when it stings, is one of the clearest ways to show you care about someone's success. Avoiding it might feel easier in the moment, but it only causes stress later. I learned that people notice. They wonder when their leader will step up and be honest. Holding back felt like a relief at first, but it meant I was dying a thousand small deaths by not saying what needed to be said.

From that point on, I made a promise to myself and my teams that I would never hold back again. I learned that constructive feedback is the fastest path to growth and trust. Everyone benefits from it. I will always be grateful to that colleague for giving me that hard truth. It set me on a different path that has fueled success for myself and for every team I have led since. It energized me, and it still does.

## A Simple Self-Assessment to Get Started

Use this quick assessment to see how you are doing on key behaviors that support strong, human-centered leadership. Be honest. This is for your growth, not your scorecard.

| Attribute | Ask Yourself |
|---|---|
| Emotional presence | Do I stay calm and focused under pressure? |
| Listening | Do I really listen to understand or just wait to respond? |

| Adaptability | Do I adjust quickly when facts change? |
|---|---|
| Curiosity | Do I actively explore new ideas? |
| Accountability | Do I own outcomes, both good and bad? |
| Resilience | Do I bounce back and learn from setbacks? |
| Influence | Do I shape decisions through credibility, not just authority? |
| Systems thinking | Do I see how my work connects across the business? |
| Ethical empathy | Do I think about the human impact of my decisions? |
| Knowledge sharing | Do I help others learn and grow by sharing what I know? |

**Action:** Choose one or two areas to focus on in the coming weeks. Make notes. Ask for feedback. Notice how small shifts in your self-awareness start to change your impact on others.

## What You Can Do Now

- **Set aside time for reflection:** Even five minutes at the end of the day makes a difference.

- **Use a journal:** Capture thoughts, lessons, and situations that did not go well so you can review them.

- **Ask someone you trust where they see your blind spots:** Listen without defending.

- **Pick one area from the assessment and work on it:** Small, focused improvements add up over time.

- **Treat yourself with respect:** Growth is a long game. Be as patient with yourself as you are with your best team members.

## The Takeaway

Self-reflection is the work that makes everything else more effective. It strengthens how leaders show up and drive results. The clearer you are about yourself, the better you will lead others. In the end, your willingness to look inward becomes the foundation that supports every team you build and every decision you make.

# CHAPTER TWENTY-TWO

## Legacy in Motion and the Real Deliverable

### Your Legacy Is Happening Now

**MOST PEOPLE THINK** of their leadership legacy as something that becomes clear after they retire or move on. The truth is your legacy is already in motion. It is how people experience you today. It shows up in the daily moments where you build trust, shape culture, and grow people.

You see it in how your team handles setbacks when you are not around, how they hold each other accountable, and how they carry your values into their own decisions. Legacy is not what you leave behind at the end. It is what you are building every day.

### Trust Is Your Best Deliverable

Trust is the most strategic thing you can deliver. Trust ensures your influence lasts. Trust speeds up decisions, lets people take smart risks, and carries your impact far beyond what technology alone can do. Trust is also what sticks. Systems change. Tools evolve. But the trust you build and the culture of accountability you create keep paying off long after your name is off the org chart.

# Three Types of Legacy

When you look at your leadership, think about it in three parts.

1. **People** – Who have you helped grow? Who is more capable or confident because of your coaching, belief, or honest feedback? Your real legacy is who you equip to lead after you.

Ask yourself:

- Have I invested in people, not just results?
- Have I taken the time to develop others?
- Have I multiplied my influence through those I have mentored?

2. **Culture** – What happens when you are not in the room? Do people act on shared values and clear expectations, or do they wait to be told? *Culture* is defined as people's behavior when no one is watching. If you have led well, it shows up in respect, accountability, and shared purpose that lasts without you.

3. **Capability** – What keeps growing after you move on? Have you put in place frameworks, decision habits, and ways of thinking that do not rely on you being there? With true capability building, the organization continues to evolve and deliver value even when you step back.

## Make Space for Others to Rise

Part of your legacy is preparing others to lead. Create chances for people to learn by doing. Delegate with purpose, letting them figure things out, and coaching them through tough moments. Your team grows most when they see you trust them. When they know you will back them up, they become more resilient and ready for what comes next.

## Teach Your Frameworks

Sharing your tools and thought processes is part of your legacy. When you show people how to evaluate investments, shape demand, or build trust, you extend your influence beyond your direct reach. The organization becomes stronger because more people know how to think that way on their own.

## Build Trust into the Culture

The goal is to create trust and accountability that exist without you. When you do this well, people do not rely on you to enforce standards. They hold each other accountable. They make decisions with customers and long-term goals in mind because that is simply how the team operates.

## Questions to Reflect On

- What would your peers say about your leadership presence?
- Who has grown because of your influence?
- Have you built your legacy on people and culture, not just personal wins?

- Is trust strong enough that it continues without you?
- Who have you prepared to lead next?

## What Truly Lasts

Your success is tied to trust, clarity, accountability, and how well you lead people through change. The systems you build matter, but they are not your greatest deliverable. The most strategic thing you offer is trusted relationships and an environment that grows people and culture. This is what keeps paying dividends long after the technology shifts.

## The Takeaway

Your legacy is already happening. It is not something that waits until the end. It lives in how people experience you now, who you have helped grow, the culture you have shaped, and the trust you have built. That is what continues to deliver value long after you are gone. It is why people will carry your influence forward and apply what they learned from you wherever they go next. Ask yourself the simplest question of all: **Will you be missed when you are gone?** The answer tells you more about your real legacy than any project, system, or personal milestone ever will.

# Afterword

**THANK YOU FOR** investing your time with this book. If you've made it to this page, it tells me you care deeply about leading well; not just delivering outcomes, but growing people, shaping culture, and leaving something that lasts.

Throughout this book, we've explored how the real work of leadership is not to outrun technology but to amplify what only humans can do. Building trust. Exercising judgment. Holding healthy tension. Creating spaces where people can do the best work of their lives.

Technology will keep getting smarter. AI will get faster. But people will always look to you for clarity, encouragement, and accountability. That's what makes your role so profoundly human. It's also why your legacy isn't something that starts after you retire. It's already in motion, in every conversation you have, every person you coach, every standard you uphold. Keep investing in relationships. Keep making the tough calls with empathy and courage. Keep building people. Don't forget to have some fun along the way. The energy you bring sets the tone for everyone else.

Thank you for letting me be part of your leadership growth in the AI world. Imagine how powerful the next chapters of your leadership can become and all the remarkable achievements you'll make possible. In the end, it is your humanity that will carry you far beyond the algorithm.

# Notes and Further Reading

**THE FOLLOWING REFERENCES** acknowledge the original sources of key concepts mentioned in this book and provide readers with resources for deeper exploration.

- **Wakeman, Cy.** *Reality-Based Leadership: Ditch the Drama, Restore Sanity to the Workplace, and Turn Excuses into Results.* Jossey-Bass, 2010.
  – Referenced in the Foreword, Chapters 7, 10.

- **Goleman, Daniel.** *Emotional Intelligence: Why It Can Matter More Than IQ.* Bantam Books, 1995.
  – Referenced in Chapter 10.

- **Karpman, Stephen.** "Fairy Tales and Script Drama Analysis." *Transactional Analysis Bulletin*, vol. 7, no. 26, 1968, pp. 39–43.
  – Referenced in Chapter 10.

# BEYOND THE ALGORITHM
# Index

*Italic* page references indicate boxed text.

www.ingramcontent.com/pod-product-compliance
Lightning Source LLC
Chambersburg PA
CBHW070530200326
41519CB00013B/2999